Reactive Search and Intelligent Optimization

OPERATIONS RESEARCH/COMPUTER SCIENCE INTERFACES

Professor Ramesh Sharda
Oklahoma State University

Prof. Dr. Stefan Voß
Universität Hamburg

Bierwirth / *Adaptive Search and the Management of Logistics Systems*
Laguna & González-Velarde / *Computing Tools for Modeling, Optimization and Simulation*
Stilman / *Linguistic Geometry: From Search to Construction*
Sakawa / *Genetic Algorithms and Fuzzy Multiobjective Optimization*
Ribeiro & Hansen / *Essays and Surveys in Metaheuristics*
Holsapple, Jacob & Rao / *Business Modelling: Multidisciplinary Approaches — Economics, Operational and Information Systems Perspectives*
Sleezer, Wentling & Cude / *Human Resource Development and Information Technology: Making Global Connections*
Voß & Woodruff / *Optimization Software Class Libraries*
Upadhyaya et al / *Mobile Computing: Implementing Pervasive Information and Communications Technologies*
Reeves & Rowe / *Genetic Algorithms — Principles and Perspectives: A Guide to GA Theory*
Bhargava & Ye / *Computational Modeling and Problem Solving In the Networked World: Interfaces in Computer Science & Operations Research*
Woodruff / *Network Interdiction and Stochastic Integer Programming*
Anandalingam & Raghavan / *Telecommunications Network Design and Management*
Laguna & Martí / *Scatter Search: Methodology and Implementations in C*
Gosavi / *Simulation-Based Optimization: Parametric Optimization Techniques and Reinforcement Learning*
Koutsoukis & Mitra / *Decision Modelling and Information Systems: The Information Value Chain*
Milano / *Constraint and Integer Programming: Toward a Unified Methodology*
Wilson & Nuzzolo / *Schedule-Based Dynamic Transit Modeling: Theory and Applications*
Golden, Raghavan & Wasil / *The Next Wave in Computing, Optimization, and Decision Technologies*
Rego & Alidaee / *Metaheuristics Optimization via Memory and Evolution: Tabu Search and Scatter Search*
Kitamura & Kuwahara / *Simulation Approaches in Transportation Analysis: Recent Advances and Challenges*
Ibaraki, Nonobe & Yagiura / *Metaheuristics: Progress as Real Problem Solvers*
Golumbic & Hartman / *Graph Theory, Combinatorics, and Algorithms: Interdisciplinary Applications*
Raghavan & Anandalingam / *Telecommunications Planning: Innovations in Pricing, Network Design and Management*
Mattfeld / *The Management of Transshipment Terminals: Decision Support for Terminal Operations in Finished Vehicle Supply Chains*
Alba & Martí / *Metaheuristic Procedures for Training Neural Networks*
Alt, Fu & Golden / *Perspectives in Operations Research: Papers in Honor of Saul Gass' 80th Birthday*
Baker et al / *Extending the Horizons: Adv. In Computing, Optimization, and Dec. Technologies*
Zeimpekis et al / *Dynamic Fleet Management: Concepts, Systems, Algorithms & Case Studies*
Doerner et al / *Metaheuristics: Progress in Complex Systems Optimization*
Goel / *Fleet Telematics: Real-time Management & Planning of Commercial Vehicle Operations*
Gondran & Minoux / *Graphs, Dioïds and Semirings: New Models and Algorithms*
Alba & Dorronsoro / *Cellular Genetic Algorithms*
Golden, Raghavan & Wasil / *The Vehicle Routing Problem: Latest Advances and New Challenges*
Raghavan, Golden & Wasil / *Telecommunications Modeling, Policy and Technology*

Roberto Battiti • Mauro Brunato • Franco Mascia

Reactive Search
and Intelligent Optimization

Springer

Roberto Battiti
Università Trento
Dip. Ingegneria e Scienza dell'Informazione
Via Sommarive, 14
38100 Trento
Italy
battiti@disi.unitn.it

Franco Mascia
Università Trento
Dip. Ingegneria e Scienza dell'Informazione
Via Sommarive, 14
38100 Trento
Italy
mascia@disi.unitn.it

Mauro Brunato
Università Trento
Dip. Ingegneria e Scienza dell'Informazione
Via Sommarive, 14
38100 Trento
Italy
brunato@disi.unitn.it

http://reactive-search.org/

ISSN: 1387-666X
ISBN: 978-1-4419-3499-4   e-ISBN: 978-0-387-09624-7
DOI: 10.1007/978-0-387-09624-7

© 2010 Springer Science+Business Media, LLC
All rights reserved. This work may not be translated or copied in whole or in part without the written permission of the publisher (Springer Science+Business Media, LLC, 233 Spring Street, New York, NY 10013, USA), except for brief excerpts in connection with reviews or scholarly analysis. Use in connection with any form of information storage and retrieval, electronic adaptation, computer software, or by similar or dissimilar methodology now known or hereafter developed is forbidden.
The use in this publication of trade names, trademarks, service marks, and similar terms, even if they are not identified as such, is not to be taken as an expression of opinion as to whether or not they are subject to proprietary rights.

Printed on acid-free paper

springer.com

Contents

1 Introduction: Machine Learning for Intelligent Optimization 1
   1.1 Parameter Tuning and Intelligent Optimization 4
   1.2 Book Outline 7

2 Reacting on the Neighborhood 9
   2.1 Local Search Based on Perturbations 9
   2.2 Learning How to Evaluate the Neighborhood 13
   2.3 Learning the Appropriate Neighborhood in Variable
       Neighborhood Search 14
   2.4 Iterated Local Search 18

3 Reacting on the Annealing Schedule 25
   3.1 Stochasticity in Local Moves and Controlled Worsening
       of Solution Values 25
   3.2 Simulated Annealing and Asymptotics 26
       3.2.1 Asymptotic Convergence Results 27
   3.3 Online Learning Strategies in Simulated Annealing 29
       3.3.1 Combinatorial Optimization Problems 30
       3.3.2 Global Optimization of Continuous Functions 33

4 Reactive Prohibitions 35
   4.1 Prohibitions for Diversification 35
       4.1.1 Forms of Prohibition-Based Search 36
       4.1.2 Dynamical Systems 37
       4.1.3 A Worked-Out Example of Fixed Tabu Search 39
       4.1.4 Relationship Between Prohibition and Diversification 39
       4.1.5 How to Escape from an Attractor 41
   4.2 Reactive Tabu Search: Self-Adjusted Prohibition Period 49
       4.2.1 The Escape Mechanism 51
       4.2.2 Applications of Reactive Tabu Search 51

	4.3	Implementation: Storing and Using the Search History 52
		4.3.1 Fast Algorithms for Using the Search History 54
		4.3.2 Persistent Dynamic Sets 54

5 Reacting on the Objective Function 59
    5.1 Dynamic Landscape Modifications to Influence Trajectories 59
        5.1.1 Adapting Noise Levels 62
        5.1.2 Guided Local Search 63
    5.2 Eliminating Plateaus by Looking Inside the Problem Structure 66
        5.2.1 Nonoblivious Local Search for SAT 66

6 Model-Based Search 69
    6.1 Models of a Problem 69
    6.2 An Example 71
    6.3 Dependent Probabilities 73
    6.4 The Cross-Entropy Model 75
    6.5 Adaptive Solution Construction with Ant Colonies 77
    6.6 Modeling Surfaces for Continuous Optimization 79

7 Supervised Learning 83
    7.1 Learning to Optimize, from Examples 83
    7.2 Techniques 84
        7.2.1 Linear Regression 84
        7.2.2 Bayesian Locally Weighted Regression 88
        7.2.3 Using Linear Functions for Classification 92
        7.2.4 Multilayer Perceptrons 94
        7.2.5 Statistical Learning Theory and Support Vector Machines .. 95
        7.2.6 Nearest Neighbor's Methods 101
    7.3 Selecting Features 102
        7.3.1 Correlation Coefficient 104
        7.3.2 Correlation Ratio 104
        7.3.3 Entropy and Mutual Information 105
    7.4 Applications 106
        7.4.1 Learning a Model of the Solver 110

8 Reinforcement Learning 117
    8.1 Reinforcement Learning Basics: Learning from a Critic 117
        8.1.1 Markov Decision Processes 118
        8.1.2 Dynamic Programming 120
        8.1.3 Approximations: Reinforcement Learning
               and Neuro-Dynamic Programming 123
    8.2 Relationships Between Reinforcement Learning
        and Optimization 125

Contents

9 Algorithm Portfolios and Restart Strategies 129
   9.1 Introduction: Portfolios and Restarts 129
   9.2 Predicting the Performance of a Portfolio from its Component
        Algorithms 130
        9.2.1 Parallel Processing 132
   9.3 Reactive Portfolios 134
   9.4 Defining an Optimal Restart Time 135
   9.5 Reactive Restarts 138

10 Racing 141
   10.1 Exploration and Exploitation of Candidate Algorithms 141
   10.2 Racing to Maximize Cumulative Reward by Interval Estimation ... 142
   10.3 Aiming at the Maximum with Threshold Ascent 144
   10.4 Racing for Off-Line Configuration of Metaheuristics 145

11 Teams of Interacting Solvers 151
   11.1 Complex Interaction and Coordination Schemes 151
   11.2 Genetic Algorithms and Evolution Strategies 152
   11.3 Intelligent and Reactive Solver Teams 156
   11.4 An Example: Gossiping Optimization 159
        11.4.1 Epidemic Communication for Optimization 160

12 Metrics, Landscapes, and Features 163
   12.1 How to Measure and Model Problem Difficulty 163
   12.2 Phase Transitions in Combinatorial Problems 164
   12.3 Empirical Models for Fitness Surfaces 165
        12.3.1 Tunable Landscapes 168
   12.4 Measuring Local Search Components: Diversification and Bias 170
        12.4.1 The Diversification–Bias Compromise (D–B Plots) 173
        12.4.2 A Conjecture: Better Algorithms are Pareto-Optimal
               in D–B Plots 175

13 Open Problems 177

References 181

Index 195

Acknowledgments

> *Considerate la vostra semenza:*
> *fatti non foste a viver come bruti,*
> *ma per seguir virtute e canoscenza.*
>
> *Li miei compagni fec'io sì aguti,*
> *con questa orazion picciola, al cammino,*
> *che a pena poscia li avrei ritenuti;*
>
> *e volta nostra poppa nel mattino,*
> *de' remi facemmo ali al folle volo,*
> *sempre acquistando dal lato mancino.*
>
> *Consider your origins:*
> *you're not made to live as beasts,*
> *but to follow virtue and knowledge.*
>
> *My companions I made so eager,*
> *with this little oration, of the voyage,*
> *that I could have hardly then contained them;*
>
> *that morning we turned our vessel,*
> *our oars we made into wings for the foolish flight,*
> *always gaining ground toward the left.*
>
> (Dante, Inferno Canto XXVI, translated by Anthony LaPorta)

A book is always, at least in part, a collective intelligence creation, and it is always difficult to acknowledge the countless comments and contributions obtained by different colleagues, friends, and students, while reading preliminary versions, commenting, and discovering mistakes. We thank them all and of course we keep responsibility for the remaining errors. For example, comments on different chapters have been submitted by Holger Hoos, Gabriela Ochoa, Vittorio Maniezzo, Fred Glover, Matteo Gagliolo, Qingfu Zhang, Mark Jelasity, Frank Hutter, Nenad Mladenović, Lyle McGeoch, Christian Blum, Stefan Voss, Alberto Montresor, Nicolò Cesa-Bianchi, Jean-Paul Watson, Michail Lagoudakis, Christine Solnon, and Francesco Masulli. In particular, we thank Elisa Cilia, Paolo Campigotto, and our young LION team members Roberto Zandonati, Marco Cattani, Stefano Teso, Cristina Di Risio, Davide Mottin, and Sanjay Rawat for their fresh initial look at the book topics and their constructive comments. Finally, some of the figures in this book are courtesy of the artistic ability of Marco Dianti (Figs. 1.1, 1.4, and 2.1), and Themis Palpanas (Fig. 2.2).

Last but not least, we thank our partners (Annamaria, Anna, and Elisa) for their patience during the most intense moments related to writing this book.

We are happy to hear from our readers, for signaling opinions, suggesting changes, impressions, novel developments. Please use one of the following emails: `battiti@disi.unitn.it`, or `brunato` or `mascia`. The reactive search online community at

> `http://reactive-search.org/`
> `http://reactive-search.com/`

and the LION laboratory (machine Learning and Intelligent OptimizatioN) Web site at

> `http://intelligent-optimization.org/`

are good places to look for updated information, software, and teaching material related to this book.

Roberto, Mauro, and Franco

Chapter 1
Introduction: Machine Learning for Intelligent Optimization

> *Errando discitur.*
> *Chi fa falla; e fallando s'impara.*
> *We learn from our mistakes.*
> *(Popular wisdom)*
>
> *You win only if you aren't afraid to lose.*
> *(Rocky Aoki)*
>
> *Mistakes are the portals of discovery.*
> *(James Joyce)*

This book is about learning for problem solving. If you are not already an expert in the area, let us start with some motivation, just to make sure that we talk about issues that are not far from everybody's human experience.

Human problem solving is strongly connected to learning. Learning takes place when the problem at hand is not well known at the beginning, and its structure becomes clearer and clearer when more experience with the problem is available. For concreteness, let us consider skiing. What distinguishes an expert skier from a novice is that the novice knows some instructions but needs a lot of experience to *fine tune* the techniques (with some falling down into local minima and restarts), while the real expert jumps *seamlessly* from sensorial input to action, without effort and reasoning. The knowledge accumulated from the previous experience has been *compiled into parameters of a neural system* working at very high speed. Think about yourself driving a car, and try to explain in detail how you move your feet when driving: After so many years the knowledge is so hardwired into your neural system that you hardly need any high-level thinking. Of course, this kind of fine tuning of strategies and knowledge compilation into parameters of a dynamical system (our nervous system) is quite natural for us, while more primitive creatures are more rigid in their behavior. Consider a fly getting burnt by an incandescent light bulb. It is deceived by the novel situation because no light bulb was present during the genetic evolution of its species, apart from a very distant one called "sun." You know the rest of the story: The fly will get burnt again and again and again (Fig. 1.1). Lack of lifelong learning and rapid self-tuning can have disastrous consequences. In fact, the human ability of learning and quick reaction to life-threatening dangers and novel contexts has been precious for the survival of our species when our ancestors were living in unfriendly environments such as forests, hunted by animals with higher physical strength. In addition to learning, search by trial-and-error, generation, and test, repeated modifications of solutions by small local changes are also part of human life.

Fig. 1.1 Attracted by the lightbulb and repeatedly burnt: The fly is not intelligent enough to learn from experience and develop an escape strategy. (Image courtesy of Marco Dianti)

What is critical for humans is also critical for many human-developed problem-solving strategies. It is not surprising that many methods for solving problems in artificial intelligence, operations research, and related areas follow the *search* scheme, for example, *searching for an optimal configuration on a tree of possibilities* by adding one solution component at a time, and backtracking if a dead-end is encountered, or *searching by generating a trajectory of candidate solutions* on a landscape defined by the corresponding solution value.

For most of the relevant and difficult problems (look for "computational complexity" and "NP-hardness" in your favorite search engine), researchers now believe that *the optimal solution cannot be found exactly in acceptable computing times*, which grow as a low-order polynomial of the input size. In some cases, similar negative results exist about the possibility of approximating a solution with performance guarantees (*hardness of approximation*). These are well-known negative results established in the last decades of theoretical computer science. Hardness of approximation, in addition to NP-hardness, is a kind of "affirmative action" for heuristics.

Heuristics used to suffer from a bad reputation, citing from Papadimitriou and Steiglitz [205]:

> 6. *Heuristics* Any of the <five> approaches above without a formal guarantee of performance can be considered a "heuristic." However *unsatisfactory mathematically*, such approaches are certainly valid in practical situations.

Unfortunately, because of the above-mentioned theoretical results, we are condemned to live with heuristics for very long times, maybe forever, and some effort is required to make them more satisfactory, both from a theoretical and from a practical point of view. This implies that an important part of computer science is becoming experimental: some of its theories cannot be demonstrated by deduction as in mathematics, but have to be subjected to careful experimentation. Terms such as *experimental algorithmics* and *experimental computer science* have been used to underline this transformation. To make an analogy: There is no mathematical proof of Newton's law of gravitation, it simply describes in a *unified* manner many related physical phenomena, apples falling from a tree and planets orbiting around other planets, and it has been refined and subjected to careful experimentation.

In addition, there is no mathematical proof that the speed of light is independent of the observer, in fact a careful experiment was needed to demonstrate that this is indeed the case. Similarly, serious experimental work can be done in computer science, actually *it must be done* given the mentioned theoretical difficulties in proving results, without being accused of lack of mathematical proofs.

A particular delicate issue in many heuristics is their detailed tuning. Some schemes are not rigid and demand the specification of choices in the detailed algorithm design, or values of internal parameters. Think about our novice skier, its detailed implementation ranging from world champion to ... the writers of this book: The difference in parameter tuning is evident.

Parameter tuning is a crucial issue both in the scientific development and in the practical use of heuristics. In some cases the detailed tuning is executed by a researcher or by a final user. As parameter tuning is user dependent, the reproducibility of the heuristics results is difficult as is comparing different parametric algorithms. Algorithm A can be better than algorithm B if tuned by Roberto, while it can be worse if tuned by Mauro.

In this book we consider some machine learning methods that can be profitably used in order to *automate the tuning* process and make it an integral and *fully documented* part of the algorithm. In particular, the focus is on learning schemes where the accumulated knowledge is compiled into the parameters of the method, or the parameters regulating a dynamical system to search for a solution. These schemes are called *subsymbolic* to differentiate them from high-level reasoning schemes popular in artificial intelligence. In many cases subsymbolic learning schemes work without giving a high-level symbolic explanation. Think about neural networks, and about the champion skier who cannot explain how he allocates forces to the different muscles during a slalom competition.

If learning acts *online*, i.e., while the algorithm is solving an instance of a problem, *task-dependent local properties* can be used by the algorithm to determine the appropriate balance between *diversification* and *intensification*. Deciding whether it is better to look for gold where the other miners are excavating (*intensification/exploitation*) or to go and explore other valleys and uncharted territories (*diversification/exploration*) is an issue that excruciated forty-niners, which we will meet over and over again in the following chapters. Citing for example from [217], "diversification drives the search to examine new regions, and intensification focuses more intently on regions previously found to be good. Intensification typically operates by re-starting from high quality solutions, or by modifying choice rules to favor the inclusion of attributes of these solutions."

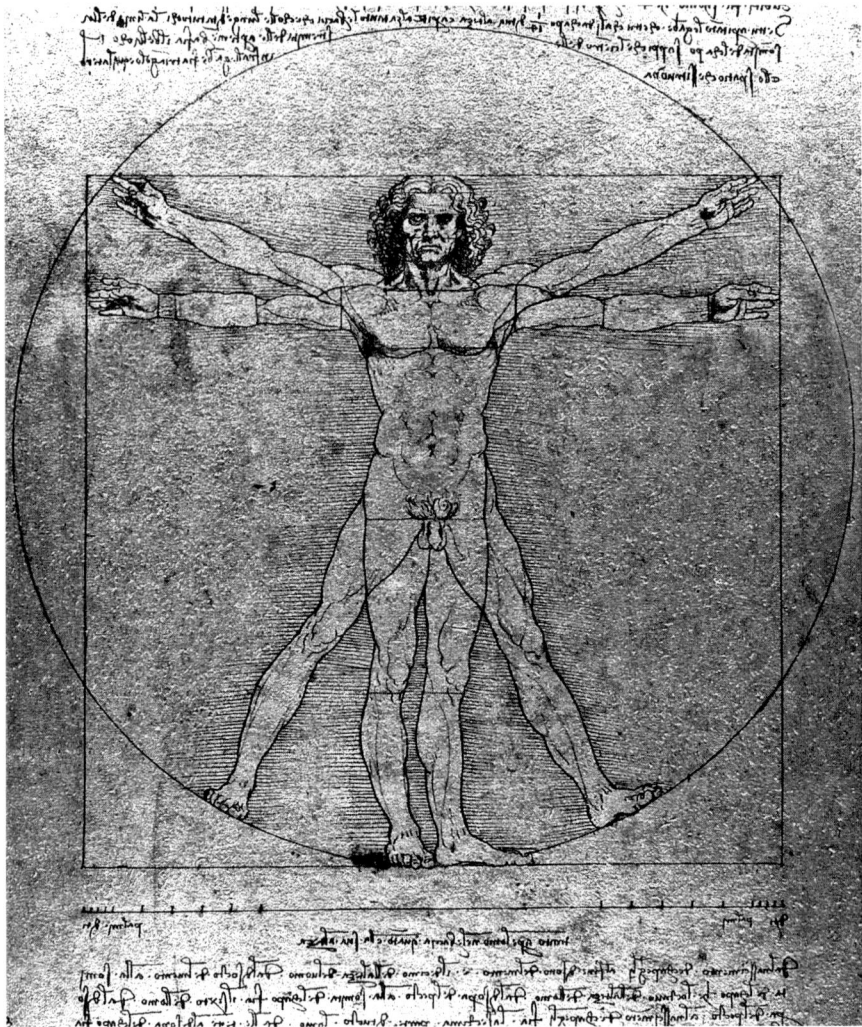

Fig. 1.2 Who invented reactive search? Man is the best example of an effective online learning machine (Leonardo Da Vinci, *Vitruvian Man*, Gallerie dell'Accademia, Venice)

1.1 Parameter Tuning and Intelligent Optimization

As we mentioned, many state-of-the-art heuristics are characterized by a certain number of *choices and free parameters*, whose appropriate setting is a subject that raises issues of research methodology [15, 135, 182]. In some cases the parameters are tuned through a feedback loop that includes *the user as a crucial learning component*: Different options are developed and tested until acceptable results are obtained. The quality of results is not automatically transferred to different instances

1.1 Parameter Tuning and Intelligent Optimization

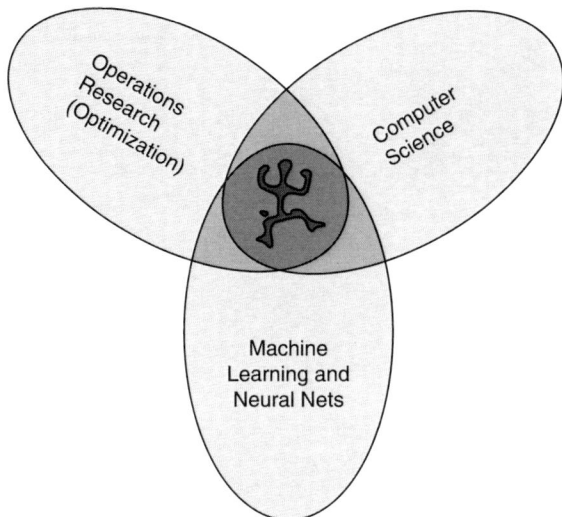

Fig. 1.3 Machine learning and intelligent optimization as the intersection of three disciplines. The logo derived from cave art is related to the first attempts of men to describe themselves in action

and the feedback loop can require a slow "trial and error" process when the algorithm has to be tuned for a specific application. The machine learning community, with significant influx from statistics, developed in the last decades a rich variety of "design principles" that can be used to develop machine learning methods that can be profitably used in the area of parameter tuning for heuristics. In this way one *eliminates the human intervention*. This does not imply higher unemployment rates for researchers. On the contrary, one is now loaded with a heavier task: The algorithm developer must transfer his intelligent expertise into the algorithm itself, a task that requires the exhaustive description of the tuning phase *in the algorithm*. The algorithm complication will increase, but the price is worth paying if the two following objectives are reached:

- *Complete and unambiguous documentation.* The algorithm becomes self-contained and its quality can be judged independently from the designer or specific user. This requirement is particularly important from the scientific point of view, where objective evaluations are crucial. The recent introduction of software archives further simplifies the test and *simple reuse* of heuristic algorithms.
- *Automation.* The time-consuming handmade tuning phase is now substituted by an automated process. Let us note that only the final user will typically benefit from an automated tuning process. On the contrary, the algorithm designer faces a longer and harder development phase, with a possible preliminary phase of exploratory tests, followed by the above described exhaustive documentation of the tuning process when the algorithm is presented to the scientific community.

Reactive search advocates the integration of subsymbolic machine learning techniques into search heuristics for solving complex optimization problems. The word

Fig. 1.4 If algorithms have self-tuning capabilities, complex problem solving does not require technical expertise but is available to a much wider community of final users. (Image courtesy of Marco Dianti)

reactive hints at a ready response to events during the search through an internal *feedback loop for online self-tuning and dynamic adaptation*. In reactive search the past history of the search and the knowledge accumulated while moving in the configuration space is used for self-adaptation in an autonomic manner: The algorithm maintains the internal flexibility needed to address different situations during the search, but the adaptation is automated, and executed while the algorithm runs on a single instance and reflects on its past experience (Fig. 1.2).

Methodologies of interest for reactive search include machine learning and statistics, in particular reinforcement learning and neuro-dynamic programming, active or query learning, and neural networks.

Intelligent optimization, a superset of reactive search, refers to a more extended area of research, including online and offline schemes based on the use of memory, adaptation, incremental development of models, experimental algorithmics applied to optimization, intelligent tuning, and design of heuristics. In some cases the work is at an upper level, where basic methods are properly guided and combined, and the term *metaheuristics* has been proposed in the past. A problem with this term is that the boundary signaled by the "meta" prefix is not always clear: In some cases the intelligence is embedded and the term *intelligent optimization* underlines the holistic point of view concerned with complete systems rather than with their dissection into parts (the basic components and the meta-level) (Fig. 1.3).

The metaphors for reactive search derive mostly from the individual human experience. Its motto can be "learning on the job." Real-world problems have a rich structure. While many alternative solutions are tested in the exploration of a search space, patterns and regularities appear. The human brain quickly learns and drives future decisions based on previous observations. This is the main inspiration source for inserting online machine learning techniques into the optimization engine of reactive search. *Memetic* algorithms share a similar focus on learning, although their concentration is on *cultural evolution*, describing how societies develop over time, more than on the capabilities of a single individual.

Nature and biology-inspired metaphors for optimization algorithms abound in this time frame. It is to some degree surprising that most of them derive from genetics and evolution, or from the emergence of collective behaviors from the interaction of simple living organisms that are mostly hard-wired with little or no learning capabilities. One almost wonders whether this is related to ideological prejudices in the spirit of Jean-Jacques Rousseau, who believed that man was good when in the state of nature but is corrupted by society, or in the spirit of the "evil man against nature." But metaphors lead us astray from our main path: we are strong supporters of a pragmatic approach, an algorithm is effective if it solves a problem in a competitive manner without requiring an expensive tailoring, not because it corresponds to one's favorite elaborate, fanciful, or sexy analogies. Furthermore, at least for a researcher, in most cases an algorithm is of scientific interest if there are ways to analyze its behavior and explain why and when it is effective.

As a final curiosity, let us note that the term *reactive* as "readily responsive to a stimulus" used in our context is not in contrast with proactive as "acting in anticipation of future problems, needs, or changes." In fact, to obtain a reactive algorithm, the designer needs to be proactive by appropriately planning modules into the algorithm so that its capability of autonomous response increases. In other words, reactive algorithms need proactive designers!

1.2 Book Outline

The book does not aim at a complete coverage of the widely expanding research area of heuristics, metaheuristics, stochastic local search, etc. The task would be daunting and bordering on the myth of Sisyphus, condemned by the gods to ceaselessly rolling a rock to the top of a mountain, whence the stone would fall back of its own weight. The rolling stones are in this case caused by the rapid development of new heuristics for many problems, which would render a book obsolete after a short span.

We aim at giving the main principles and at developing some fresh intuition for the approaches. We like mathematics but we also think that hiding the underlying motivations and sources of inspiration takes some color out of the scientific work

("Grau, teurer Freund, ist alle Theorie. Und grün des Lebens goldner Baum" – Gray, my good friend, is all theory, and green is life's own golden tree – Johann Wolfgang von Goethe). On the other hand, pictures and hand-waving can be very dangerous in isolation and we try to avoid these pitfalls by also giving the basic equations when possible, or by at least directing the reader to the bibliographic references for deepening a topic.

The point of view of the book is to look at the zoo of different optimization beasts to underline *opportunities for learning and self-tuning strategies*. A leitmotiv is that seemingly diverse approaches and techniques, when observed from a sufficiently abstract point of view, show the deep interrelationships and *unity*, which is characteristic of science.

The focus is definitely more on methods than on problems. We introduce some problems to make the discussion more concrete or when a specific problem has been widely studied by reactive search and intelligent optimization heuristics.

Intelligent optimization, the application of machine learning strategies in heuristics is a very wide area, and the space in this book dedicated to *reactive search* techniques (online learning techniques applied to search methods) is wider because of personal interest. This book is mainly dedicated to local search and variations, although similar reactive principles can be applied also for other search techniques (for example, tree search).

The structure of most of the following chapters is as follows: (i) the basic issues and algorithms are introduced, (ii) the parameters critical for the success of the different methods are identified, and (iii) opportunities and schemes for the automated tuning of these parameters are presented and discussed.

Let us hit the road.

Chapter 2
Reacting on the Neighborhood

> *How many shoe-shops should one person visit before making a choice?*
> *There is not a single answer, please specify whether male or female!*
> (An author of this book who prefers anonymity)

2.1 Local Search Based on Perturbations

A basic problem-solving strategy consists of starting from an initial tentative solution and trying to improve it through repeated small changes. At each repetition the current configuration is slightly modified (*perturbed*), the function to be optimized is tested, the change is kept if the new solution is better, otherwise another change is tried. The function $f(X)$ to be optimized is called with more poetic names in some communities: *fitness* function, *goodness* function, *objective* function.

Figure 2.1 shows an example in the history of bike design. Do not expect historical fidelity here; this book is about algorithms! The first model is a starting solution with a single wheel; it works but it is not optimal yet. The second model is a randomized attempt to add some pieces to the original designs, the situation is worse. One could revert back to the initial model and start other changes. But let us note that, if one insists and proceeds with a second addition, one may end up with the third model, clearly superior from a usability and safety point of view! This real-life story has a lesson: Local search by small perturbations is a tasty ingredient but additional spices are in certain cases needed to obtain superior results. Let us note in passing that everybody's life is an example of an optimization algorithm in action: most of the changes are localized; dramatic changes do happen, but not so frequently. The punctilious reader may notice that the goodness function of our life is not so clearly defined. To this we answer that this book is not about philosophy, let us stop here with far-fetched analogies and go down to the nitty-gritty of the algorithms.

Local search based on perturbing a candidate solution is a first paradigmatic case where simple learning strategies can be applied. Let us define the notation. $\mathscr{X}$ is the search space, $X^{(t)}$ is the current solution at iteration ("time") t. $N(X^{(t)})$ is the neighborhood of point $X^{(t)}$, obtained by applying a set of basic moves $\mu_0, \mu_1, ..., \mu_M$ to the current configuration:

$$N(X^{(t)}) = \{X \in \mathscr{X} \text{ such that } X = \mu_i(X^{(t)}), i = 0,...,M\}$$

Fig. 2.1 A local search example: How to build a better bike, from the initial model (*left*) to a worse variation (*middle*), to the final and better configuration (*right*). (Image courtesy of Marco Dianti)

If the search space is given by binary strings with a given length L: $\mathscr{X} = \{0,1\}^L$, the moves can be those changing (or complementing or *flipping*) the individual bits, and therefore M is equal to the string length L.

Local search starts from an admissible configuration $X^{(0)}$ and builds a *search trajectory* $X^{(0)},...,X^{(t+1)}$. The successor of the current point is a point in the neighborhood with a lower value of the function f to be minimized. If no neighbor has this property, i.e., if the configuration is a local minimizer, the search stops. Let us note that maximization of a function f is the same problem as minimization of $-f$. Like all symmetric situation, this one can create some confusion with the terminology. For example, steepest descent assumes a minimizing point of view, while hill climbing assumes the opposite point of view. In most of the book we will base the discussion on minimization, and *local minima* will be the points that cannot be improved by moving to one of their neighbors. *Local optimum* is a term that can be used both for maximization and minimization.

$$Y \leftarrow \text{Improving-Neighbor}(N(X^{(t)})) \tag{2.1}$$

$$X^{(t+1)} = \begin{cases} Y & \text{if } f(Y) < f(X^{(t)}) \\ X^{(t)} & \text{otherwise (search stops)} \end{cases} \tag{2.2}$$

IMPROVING-NEIGHBOR returns an improving element in the neighborhood. In a simple case, this is the element with the lowest f value, but other possibilities exist, as we will see in what follows.

Local search is surprisingly effective because most combinatorial optimization problems have a very *rich internal structure* relating the configuration X and the f value. The analogy when the input domain is given by real numbers in $\mathbb{R}^n$ is that of a continuously differentiable function $f(x)$ – continuous with continuous derivatives. If one stays in the neighborhood, the change is limited by the maximum value of the derivative multiplied by the displacement. In more dimensions,

2.1 Local Search Based on Perturbations

the vector containing the partial derivatives is the gradient, and the change of f after a small displacement is approximated by the scalar product between the gradient and the displacement. This is why moving along the direction of the negative gradient causes the fastest possible decrease per unit of displacement, motivating the term *steepest descent*, or *gradient descent* for the basic minimization strategy. By the way, steepest descent is not necessarily the best method to adopt when one considers global convergence [18]: Greediness does not always win, both in the discrete and in the continuous domain.

Combinatorial optimization needs different measures to quantify the notion that a small change of the configuration is coupled, at least in a statistical way, to a small change of f, see also Chap. 12 about metrics.

Now, a neighborhood is suitable for local search if it reflects the problem structure. For example, if the solution is given by a permutation (like in the Traveling Salesman Problem –TSP– or in sorting), an improper neighborhood choice would be to consider single-bit changes of a binary string describing the current solution, which would immediately cause illegal configurations, not corresponding to encodings of permutations. A better neighborhood can be given by all transpositions that exchange two elements and keep all others fixed. In general, a sanity check for a neighborhood controls if the f values in the neighborhood are correlated to the f value of the current point. If one starts at a good solution, solutions of similar quality can, on the average, be found more *in its neighborhood* than by sampling a completely unrelated random point. By the way, sampling a random point generally is much more expensive than sampling a neighbor, provided that the f value of the neighbors can be updated ("incremental evaluation") and it does not have to be recalculated from scratch.

For many optimization problems of interest, a closer approximation to the global optimum is required, and therefore more complex schemes are needed in order to continue the investigation into new parts of the search space, i.e., to *diversify* the search and encourage *exploration*. Here a second structural element comes to the rescue, related to the overall distribution of local minima and corresponding f values. In many relevant problems, local minima tend to be *clustered*; furthermore, good local minima tend to be closer to other good minima. *Local minima like to be in good company!* Let us define as *attraction basin* associated to a local optimum the set of points X that are mapped to the given local optimum by the local search trajectory. An hydraulic analogy, where the local search trajectory is now the trajectory of drops of water pulled by gravity, is that of *watersheds*, regions bounded peripherally by a divide and draining ultimately to a particular lake.

Now, if local search stops at a local minimum, *kicking* the system to a close attraction basin can be much more effective than restarting from a random configuration. If evaluations of f are incremental, completing a sequence of steps to move to a nearby basin can also be *much faster* than restarting with a complete evaluation followed by a possibly long trajectory descending to another local optimum. Try to move from the bottom of the Grand Canyon to the Death Valley if not convinced.

Fig. 2.2 Two local searchers on the mountains surrounding Trento, Italy. (Photo courtesy of Themis Palpanas)

This structural property is also called *big valley*, see Fig. 2.2. For example, the work in [258] presents a probabilistic analysis of local minima distribution in the TSP; see also Chap. 12 about metrics for additional details.

To help the intuition, one may think about a smooth f surface in a continuous environment, with basins of attraction that tend to have a nested, "fractal" structure. According to Mandelbrot, a fractal is generally "a rough or fragmented geometric shape that can be subdivided into parts, each of which is (at least approximately) a reduced-size copy of the whole," a property called self-similarity. The term derives from the Latin *fractus* meaning "broken" or "fractured."

A second continuous analogy is that of a (periodic) function containing components at different wavelengths when analyzed with a Fourier transform. If you are not expert in Fourier transforms, think about looking at a figure with defocusing lenses. At first the large-scale details will be revealed – for example, a figure of a distant person; then by focusing, finer and finer details will be revealed: face, arms and legs, then fingers, hair, etc. The same analogy holds for music diffused by loudspeakers of different quality, allowing higher and higher frequencies to be heard. At each scale the sound is not random noise and a pattern, a nontrivial structure is always present, see Fig. 2.3. This multiscale structure, where smaller valleys are nested within larger ones, is the basic motivation for methods such as Variable Neighborhood Search (VNS), see Sect. 2.3, and Iterated Local Search (ILS), see Sect. 2.4.

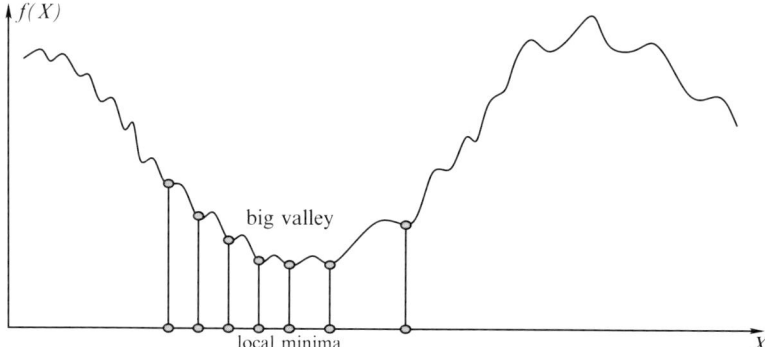

Fig. 2.3 Structure in optimization problems: The "big valley" hypothesis

2.2 Learning How to Evaluate the Neighborhood

It looks like there is little online or offline learning to be considered for a simple technique such as local search. Nonetheless, a closer look reveals some possibilities. In the function IMPROVING-NEIGHBOR one has to decide about a *neighborhood* (a set of local moves to be applied) and about a *way to pick one of the neighbors* to be the next point along the search trajectory. It is well known that selecting a neighborhood structure appropriate to a given problem is the most critical issue in the development of efficient strategies. This choice is mostly related to offline learning techniques, like those used formally or informally by researchers during the algorithm design process. But let us concentrate on *online* learning strategies that can be applied *while* local search runs on a specific instance. They can be applied in two contexts: selection of the neighbor or selection of the neighborhood. A third option is of course to adapt both choices during a run.

Let us start from the first context where a neighborhood is chosen before the run is started, and only the process to select an improving neighbor is dynamic during the run. The average progress in the optimization per unit of computational effort (the average "speed of descent" Δf_{best} per second) will depend on two factors: the average *improvement per move* and the average *CPU time per move*. There is a trade-off here: The longer to evaluate the neighborhood, the better the chance of identifying a move with a large improvement, but the shorter the total number of moves that one can execute in the CPU time allotted. The optimal setting depends on the problem, the specific instance, and the local configuration of the f landscape.

The immediate brute-force approach consists of considering *all neighbors*, by applying all possible basic moves, evaluating the corresponding f values and moving to the neighbor with the best value, breaking ties randomly if they occur. The best possible neighbor is chosen at each step. To underline this fact, the term *best-improvement local search* is used.

A second possibility consists of evaluating *a sample of the possible moves*, a subset of neighbors. In this case, IMPROVING-NEIGHBOR can return the first candidate with a better f value. This option is called FIRSTMOVE. If no such candidate exists, the trajectory is at a local optimum. A randomized examination order can be used to avoid spurious effects because of the specific examination order. FIRSTMOVE is clearly adaptive: The exact number of neighbors evaluated before deciding the next move depends not only on the instance but on the particular local properties in the configuration space around the current point. One may expect that a small number of candidates needs to be evaluated in the early phase of the search, whereas identifying an improving move will become more and more difficult during the later phases, when the configuration will be close to local optimality. The analogy is that of learning a new language: The progress is fast at the beginning but it gets slower and slower after reaching an advanced level. To repeat, if the number of examined neighbors is adapted to the evaluation results (keep evaluating until either an improving neighbor is found or all neighbors have been examined), no user intervention is needed for the self-tuning of the appropriate number of neighbors.

Another possibility consists of examining a *random subset* of neighbors, while ensuring that the sample is representative of the entire neighborhood; for example, stopping the examination when the possibility of finding better improving values is not worth the additional computational effort.

A radical proposal that avoids analyzing any neighborhood and that chooses a neighbor according to *utility values determined by reinforcement learning* is presented in [196]. The neighbor choice is dictated by the best-utility one among a set of *repair* heuristics associated to constraints in constraint satisfaction problems. The purpose is to "switch between different heuristics during search in order to adapt to specific regions of the search space."

In the next section we will consider more structured strategies where the neighborhood is not fixed at the beginning and the appropriate neighborhood to use at a given iteration is picked *from a set of different neighborhoods*.

2.3 Learning the Appropriate Neighborhood in Variable Neighborhood Search

Consider the three possible sources of information to adapt a proper neighborhood: the problem, the specific instance, and the current position in the search space. There are cases when no optimal and fixed neighborhood is defined for a problem because of lack of information, or cases when adaptation of the neighborhood to the local configuration is beneficial.

To design an automated neighborhood tuning technique, one has to specify the amount of variability among the possible neighborhood structures. A possible way is to consider *a set of neighborhoods*, defined a priori at the beginning of the search, and then aim at using the most appropriate one during the search, as illustrated in Fig. 2.4. This is the seminal idea of the VNS technique, see [123].

Let the set of neighborhoods be $\{N_1, N_2, ..., N_{k\max}\}$. A proper VNS strategy has to deal with the following issues:

2.3 Learning the Appropriate Neighborhood

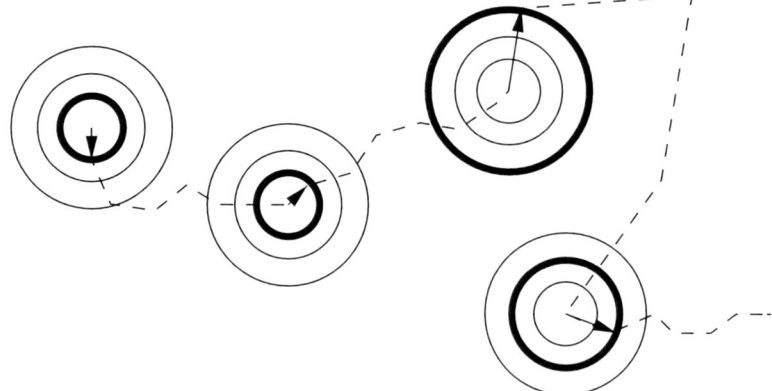

Fig. 2.4 Variable neighborhood search: The used neighborhood ("circle" around the current configuration) varies along the search trajectory

1. Which neighborhoods to use and how many of them? Larger neighborhoods may include smaller ones or be disjoint.
2. How to schedule the different neighborhoods during the search (order of consideration, transitions between different neighborhoods)?
3. Which neighborhood evaluation strategy to use (first move, best move, sampling, etc.)?

The first issue can be decided based on detailed problem knowledge, preliminary experimentation, or simply availability of off-the-shelf software routines for the efficient evaluation of a set of neighborhoods.

The second issue leads to a range of possible techniques. A simple implementation can just *cycle randomly among the different neighborhoods* during subsequent iterations: No online learning is present but possibly more robustness for solving instances with very different characteristics or for solving an instance where different portions of the search space have widely different characteristics.

Let us note that local optimality depends on the neighborhood: As soon as a local minimum is reached for a specific N_k, improving moves can in principle be found in other neighborhoods N_j, with $j \neq k$. A possibility to use online learning is based on the principle "intensification first, minimal diversification only if needed," which we often encounter in heuristics [33]. One orders the neighborhoods according to their *diameter*, or to the *strength* of the perturbation executed, or to the distance from the starting configuration to the neighbors measured with an appropriate metric. For example, if the search space is given by binary strings and the distance is the Hamming distance, one may consider as N_1 changes of a single bit, N_2 changes of two bits, etc. The default neighborhood N_1 is the one with the least diameter; if local search makes progress, one sticks to this default neighborhood. As soon as a local minimum with respect to N_1 is encountered, one tries to go to greater distances from the current point, aiming at discovering a nearby attraction basin, possibly leading to a better local optimum.

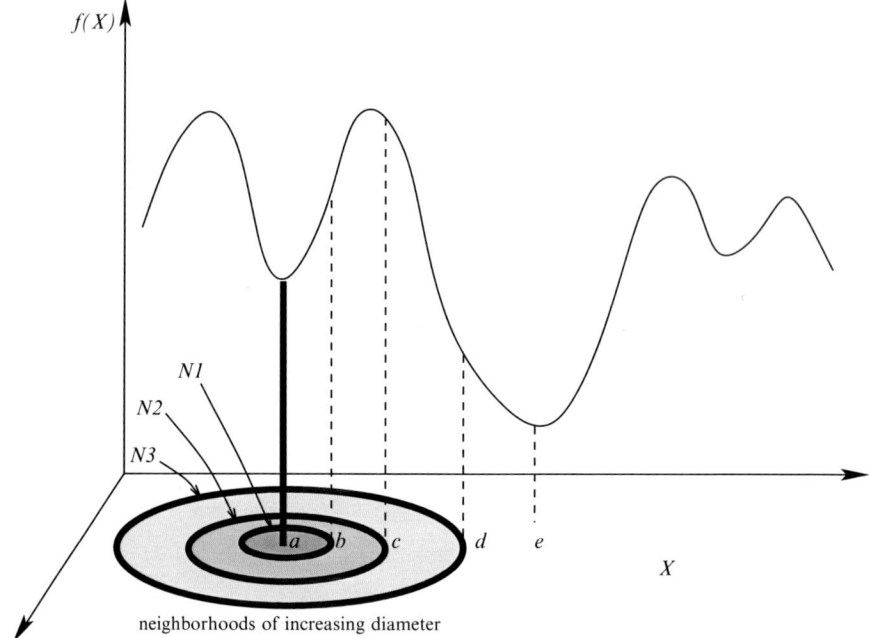

Fig. 2.5 Variable neighborhoods of different diameters. Neighboring points are on the circumferences at different distances. The figure is intended to help the intuition: The actual neighbors considered in the text are discrete

Figure 2.5 illustrates the reactive strategy: point a corresponds to the local minimum, point b is the best point in neighborhood N_1, and point c the best point in N_2. The value of point c is still worse, but the point is in a different attraction basin so that a better point e could now be reached by the default local search. The best point d in N_3 is already improving on a.

From the example one already identifies two possible strategies. In both cases one uses N_1 until a local minimum of N_1 is encountered. When this happens one considers N_2, N_3, ... In the first strategy, one stops when an improving neighbor is identified (point d in the figure). In the second strategy, one stops when one encounters a neighbor in a different attraction basin with an improving local minimum (point c in the figure). How does one know that c is in a different basin? A strategy can be that of performing a local search run from it and of looking at which point the local search converges.

For both strategies, one reverts back to the default neighborhood N_1 as soon as the *diversification* phase considering neighborhoods of increasing diameter is successful. Note a strong similarity with the design principle of Reactive Tabu Search (RTS), see Chap. 4, where diversification through prohibitions is activated when there is evidence of entrapment in an attraction basin and gradually reduced when there is evidence that a new basin has been discovered.

2.3 Learning the Appropriate Neighborhood

```
1.  function VARIABLENEIGHBORHOODDESCENT (N₁,...,N_{k_max})
2.    repeat until no improvement or max CPU time elapsed
3.        k ← 1                                              default neighborhood
4.        while k ≤ k_max:
5.            X' ← BESTNEIGHBOR (N_k(X))                     neighborhood exploration
6.            if f(X') < f(X)
7.                X ← X' ; k ← 1                             success: back to default neighborhood
8.            else
9.                k ← k+1                                    try with the following neighborhood
```

Fig. 2.6 The VND routine. Neighborhoods with higher numbers are considered only if the default neighborhood fails and only until an improving move is identified. X is the current point

```
1.  function SKEWEDVARIABLENEIGHBORHOODSEARCH (N₁,...,N_{k_max})
2.    repeat until no improvement or max CPU time elapsed
3.        k ← 1                                              default neighborhood
4.        while k ≤ k_max
5.            X'  ← RANDOMEXTRACT (N_k(X))
6.            X'' ← LOCALSEARCH(X')                          shake local search to reach local minimum
7.            if f(X'') < f(X) + αρ(X,X'')
8.                X ← X'' ; k ← 1                            success: back to default neighborhood
9.            else
10.               k ← k+1                                    try with the following neighborhood
```

Fig. 2.7 The SKEWED-VNS routine. Worsening moves are accepted provided that the change leads the trajectory sufficiently far from the starting point. X is the current point. $\rho(X,X'')$ measures the solution distance

Many VNS schemes using the set of different neighborhoods in an organized way are possible [125]. Variable Neighborhood Descent (VND), see Fig. 2.6, uses the default neighborhood first, and the ones with a higher number only if the default neighborhood fails (i.e., the current point is a local minimum for N_1), and only until an improving move is identified, after which it reverts back to N_1. When VND is coupled with an ordering of the neighborhoods according to the *strength* of the perturbation, one realizes the principle "*use the minimum strength perturbation leading to an improved solution.*"

REDUCED-VNS is a stochastic version where only one random neighbor is generated before deciding about moving or not. Line 5 of Fig. 2.6 is substituted with:

$$X' \leftarrow \text{RANDOMEXTRACT}(N_k(X))$$

SKEWED-VNS modifies the move acceptance criterion by *accepting also worsening moves if they lead the search trajectory sufficiently far* from the current point ("I am not improving but at least I keep moving without worsening too much during the diversification"); see Fig. 2.7. This version requires a suitable distance function $\rho(X,X')$ between two solutions to get controlled diversification (e.g., $\rho(X,X')$ can be the Hamming distance for binary strings), and it requires a *skewness* parameter α to regulate the trade-off between movement distance and willingness to accept worse values. By looking at Fig. 2.5, one is willing to accept the worse solution c because it is sufficiently far to possibly lead to a different attraction basin. Of course, determining an appropriate metric and skewness parameter is not a trivial task in general.

To determine an empirical α value [125], one could resort to a preliminary investigation about the distribution of local minima, by using a multistart version of VNS: One repeatedly generates random initial configurations and runs VNS to convergence. Then one studies the behavior of their expected f values as a function of the distance from the best-known solution. After collecting the above experimental data, one at least knows some reasonable ranges for α. For example, one may pick α such that the probability of success of the escape operation is different from zero. In any case, a more principled way of determined α is a topic worth additional investigation. *Valley profiles* are a second useful source of statistical information for designing an appropriate VNS strategy. They are obtained by extracting random starting points from different neighborhoods around a given point and by executing local search. Then one estimates the probabilities that the trajectories go back to the initial point or to other attraction basins and derives suggestions about the "strength of the kick," which is needed to escape with a certain probability.

Other versions of VNS employ a stochastic move acceptance criterion, in the spirit of simulated annealing as implemented in the large-step Markov-chain version [178, 180], where "kicks" of appropriate strength are used to exit from local minima; see also Sect. 2.4 about ILS.

An explicitly reactive-VNS is considered in [53] for the VRPTW (vehicle routing problem with time windows), where a construction heuristic is combined with VND using first-improvement local search. Furthermore, the objective function used by the local search operators is modified to consider the waiting time to escape from a local minimum. A preliminary investigation about a self-adaptive neighborhood ordering for VND is presented in [139]. Ratings to the different neighborhoods are derived according to their observed benefits in the past and used periodically to order the various neighborhoods.

To conclude this section, let us note some similarities between VNS and the adaptation of the search region in stochastic search technique for continuous optimization. In both cases the neighborhood is adapted to the local position in the search space. In addition to many specific algorithmic differences, let us note that the set of neighborhoods is discrete in VNS while it consists of a portion of $\mathbb{R}^n$ for continuous optimization. Neighborhood adaptation in the continuous case, see for example the Affine Shaker algorithm in [32], is mainly considered to speed-up convergence to a local minimizer, not to jump to nearby valleys.

2.4 Iterated Local Search

If a local search "building block" is available, for example as a concrete software library, how can it be used by some upper layer coordination mechanism as a black box to get better results? An answer is given by *iterating* calls to the local search rou-

2.4 Iterated Local Search

tine each time starting from a properly chosen configuration. Of course, if the starting configuration is random, one starts from scratch and knowledge about the previous searches is lost. This trivial form actually is called simply *repeated local search*.

Learning implies that the previous history, for example, the memory about the previously found local minima, is mined to produce better and better starting points. The implicit assumption is again that of a clustered distribution of local minima: Determining good local minima is easier when starting from a local minimum with a low f value than when starting from a random point. It is also faster because trajectory lengths from a local minimum to a nearby one tend to be shorter. Furthermore, an incremental evaluation of f can often be used instead of recomputation from scratch if one starts from a new point. *Updating f* values after a move can be much faster than computing them from scratch. As usual, the only caveat is to avoid confinement in a given attraction basin, so that the "kick" to transform a local minimizer into the starting point for the next run has to be appropriately strong, but not too strong to avoid reverting to memory-less random restarts (if the kick is stochastic). ILS is based on building a sequence of locally optimal solutions by: (i) perturbing the current local minimum; (ii) applying local search after starting from the modified solution.

As it happens with many simple – but sometimes very effective – ideas, the same principle has been rediscovered multiple times. For example, in [38] a local minimum of the depot location problem is perturbed by placing artificial "hazards" zones. As soon as a new point is reached, the hazard is eliminated and the usual local search starts again. Let us note that the effect of hazard zones is to penalize routes passing inside them: The f value is increased by a penalty value. In other words, the perturbation is obtained by temporarily modifying the objective function so that the old local minimum will now be suboptimal with the modified function. The first seminal papers about iterated local search and guided local search date back to $\sim$25 years ago.

One may also argue that iterated local search, at least its more advanced implementations, shares many design principles with variable neighborhood search. A similar principle is active in the iterated Lin-Kernighan algorithm of [150], where a local minimum is modified by a four-change (a "big kick" eliminating four edges and reconnecting the path) and used as a starting solution for the next run of the Lin-Kernighan heuristic. In the stochastic local search literature based on simulated annealing, the work about large-step Markov chain of [178–180] contains very interesting results coupled with a clear description of the principles.

Our description follows mainly [174]. LOCALSEARCH is seen by ILS as a black box. It takes as input an initial configuration X and ends up at a local minimum X^*. Globally, LOCALSEARCH maps from the search space $\mathscr{X}$ to the reduced set $\mathscr{X}^*$ of local minima. Obviously, the values of the objective function f at local minima are better than the values at the starting points, unless one is so lucky to start already at a local minimum. If one searches for low-cost solutions, sampling from $\mathscr{X}^*$ is therefore more effective than sampling from $\mathscr{X}$, this is in fact the basic feature of local search, see Fig. 2.8.

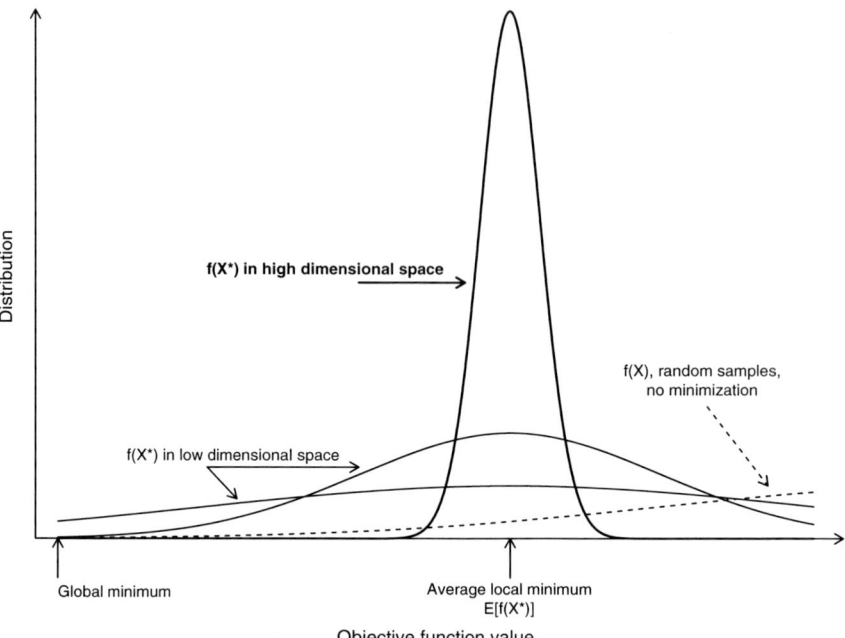

Fig. 2.8 Probability of low f values is larger for local minima in $\mathscr{X}^*$ than for a point randomly extracted from $\mathscr{X}$. Large-dimensional problems tend to have a very spiky distribution for $\mathscr{X}^*$ values

One may be tempted to sample in $\mathscr{X}^*$ by repeating runs of local search after starting from different random initial points. Unfortunately, general statistical arguments [223] related to the "law of large numbers" indicate that when the size of the instance increases, the *probability distribution of the cost values f tends to become extremely peaked about the mean* value $E[f(x^*)]$, mean value that can be offset from the best value f_{best} by a fixed percent excess. If we repeat a random extraction from $\mathscr{X}^*$, we are going to get very similar values with a large probability.

Relief comes from the rich structure of many optimization problems, which tend to cluster good local minima together. Instead of a random restart it is better to search in the neighborhood of a current good local optimum. What one needs is a *hierarchy of nested local searches*: starting from a proper neighborhood structure on $\mathscr{X}^*$ (proper as usual means that it makes the internal structure of the problem "visible" during a walk among neighbors). Hierarchy means that one uses local search to identify local minima, and then defines a local search *in the space of local minima*. One could continue, but in practice one limits the hierarchy to two levels. The sampling among $\mathscr{X}^*$ will therefore be *biased* and, if properly designed, can lead to the discovery of f values significantly lower than those expected by a random extraction in $\mathscr{X}^*$.

A neighborhood for the space of local minima $\mathscr{X}^*$, which is of theoretical interest, is obtained from the structure of the *attraction basins* around a local optimum.

2.4 Iterated Local Search

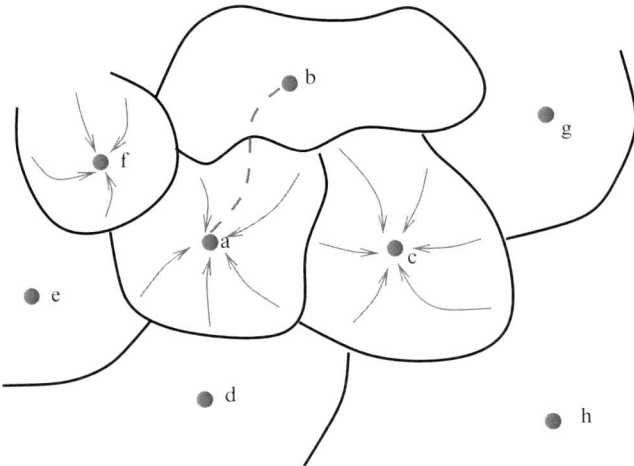

Fig. 2.9 Neighborhood among attraction basins induced a neighborhood definition on local minima in $\mathscr{X}^*$

An attraction basin contains all points that are mapped to the given optimum by local search. The local optimum is an *attractor* of the dynamical system obtained by applying the local search moves. By definition, two local minima are neighbors if and only if their attraction basins are neighbors, i.e., they share part of the boundary. For example, in Fig. 2.9, local minima b,c,d,e,f are neighbors of local minimum a. Points g,h are not neighbors of a.

A weaker notion of closeness (neighborhood) that permits a fast stochastic search in $\mathscr{X}^*$ and that does not require an exhaustive determination of the attraction basins geography – a daunting task indeed – is based on creating a *randomized path* in $\mathscr{X}$ leading from a local optimum to one of the neighboring local optima; see the path from a to b in the figure.

A final design issue is how to build the path connecting two neighboring local minima. An heuristic solution is the following one, see Figs. 2.10 and 2.11: Generate a sufficiently strong perturbation leading to a new point and then apply local search until convergence at a local minimum.

One has to determine the appropriate *strength* of the perturbation; furthermore, one has to avoid cycling: if the perturbation is too small there is the risk that the solution returns back to the starting local optimum. As a result, an endless cycle is produced if the perturbation and local search are deterministic.

Learning based on the previous search history, on memory of the previous search evolution, is of paramount importance to avoid cycles and similar traps. The principle of "intensification first, minimal diversification only if needed" can be applied, together with stochastic elements to increase robustness and discourage cycling. As we have seen for VNS, minimal perturbations maintain the trajectory in the starting attraction basin, while excessive ones bring the method closer to a random sampling, therefore loosing the boost from the problem structure properties. A possible

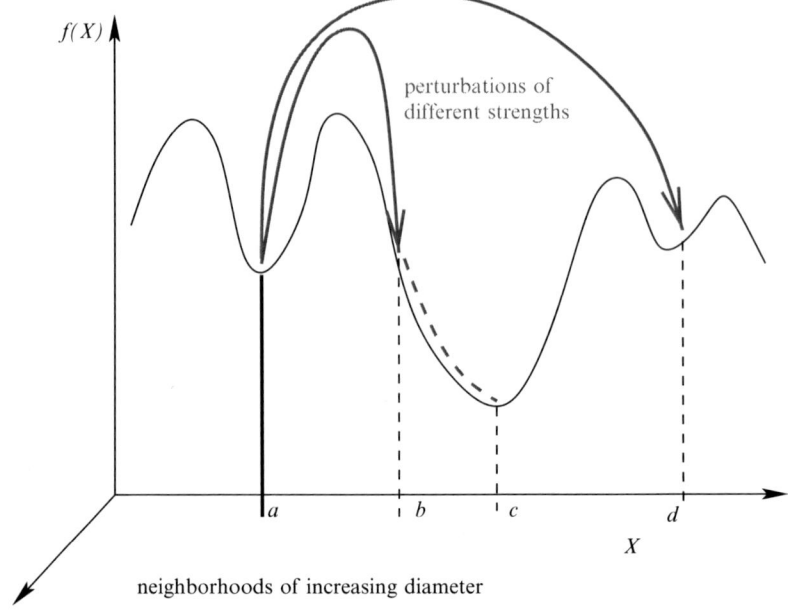

Fig. 2.10 ILS: A perturbation leads from a to b, then local search to c. If perturbation is too strong one may end up at d, therefore missing the closer local minima

```
1.  function ITERATEDLOCALSEARCH ()
2.      X⁰ ← INITIALSOLUTION()
3.      X* ← LOCALSEARCH (X⁰)
4.      repeat
5.          k ← 1                                          default neighborhood
6.          while k ≤ kmax
7.              X' ← PERTURB (X*, history)
8.              X*' ← LOCALSEARCH (X')
9.              X* ← ACCEPTANCEDECISION (X*, X*', history)
10.     until (no improvement or termination condition)
```

Fig. 2.11 Iterated Local Search

solution consists of perturbing by a short random walk of a length that is *adapted* by statistically monitoring the progress in the search.

While simple implementations of ILS are often adequate to improve on local search results, and do not require opening the "black box" local search, high performance can be obtained by *jointly optimizing the four basic components*: INITIALSOLUTION, LOCALSEARCH, PERTURB, and ACCEPTANCEDECISION. Greedy construction is often recommended in INITIALSOLUTION to identify quickly low-cost solutions, while the effect of the initial point is less relevant for long runs provided that sufficient exploration is maintained by the algorithm.

2.4 Iterated Local Search

Memory and reactive learning can be used in a way similar to that of [28] to adapt the *strength* of the perturbation to the local characteristics in the neighborhood of the current solution for the considered instance. Creative perturbations can be obtained by temporarily changing the objective function with penalties so that the current local minimum is displaced, like in [38, 72], or by *fixing* some configuration variables and by optimizing subparts of the problem [173]. The ACCEPTANCEDECISION in its different realizations can range from a strict requirement of improvement, which accepts a move only if it improves the f value, to a very relaxed *random walk*, which always accepts the randomly generated moves to favor diversification, to an intermediate *simulated annealing* test, which accepts a move depending on the f difference and on a *temperature* parameter T, with probability: $\exp\{(f(X^*) - f(X^{*\prime}))/T\}$, leading to the *large-step Markov chain* implementation of [178, 179].

It is already clear, and it will hopefully become more so in the following chapters, that the design principles underlying many superficially different techniques are in reality strongly related. We already mentioned the issue related to designing a proper perturbation, or "kick," or selecting the appropriate neighborhood, to lead a solution away from a local optimum, as well as the issue of using online reactive learning schemes to increase the robustness and permit more *hands-off* usage of software for optimization.

In particular, the simulated annealing method [161] is a popular technique to also allow worsening moves in a stochastic way. In other schemes, diversification can be obtained through the temporary *prohibition* of some local moves. Glover [107] and, independently, Hansen and Jaumard [124] popularized the Tabu Search (TS) approach. Similar ideas have been used in the past for the Traveling Salesman [241] and graph partitioning [159] problems.

We will encounter these more complex techniques in the following chapters.

Chapter 3
Reacting on the Annealing Schedule

Temperature schedules are cool.
(R. Battiti)

3.1 Stochasticity in Local Moves and Controlled Worsening of Solution Values

As mentioned previously, local search stops at a locally optimal point. Now, for problems with a rich internal structure encountered in many applications, *searching in the vicinity of good local minima* may lead to the discovery of even better solutions. In Chap. 2 one changed the neighborhood structure so as to push the trajectory away from the local minimum. In this chapter the neighborhood structure is *fixed*, but the move generation and acceptance are stochastic and one also permits a "controlled worsening" of solution values aiming at escaping from the local attractor. Let us note that a "controlled worsening" has already been considered in the SKEWED-VNS routine of Sect. 2.3, where one accepts worsening moves provided that they lead the trajectory sufficiently far.

Now, if one extends local search by *accepting worsening moves* (moves leading to worse f values) the trajectory moves to a neighbor of a local minimum. But the danger is that, after raising the solution value at the new point, the starting local minimum will be chosen at the second iteration, leading to an endless cycle of "trying to escape and falling back immediately to the starting point." This situation surely happens if the local minimum is *strict* (all neighbors have worse f values) and if more than one step is needed before points with f values better than that of the local minimum become accessible. Better points can become accessible when they can be reached from the current solution point by a local search trajectory.

The simulated annealing (SA) method has been investigated to avoid deterministic cycles and to allow for worsening moves, while still biasing the exploration so that low f values are visited more frequently than large values. The terms and the initial application comes from annealing in metallurgy, a process involving heating and controlled cooling of a metal to increase the size of its crystals and reduce their defects. The heat causes the atoms to be shaken from their initial positions, a local minimum of the internal energy, and wander randomly through states of higher energy; the slow cooling gives them more chances of finding states with lower internal energy than the initial one.

We summarize the technique, hint at mathematical results, and then underline opportunities for self-tuning.

3.2 Simulated Annealing and Asymptotics

The simulated annealing method [161] is based on the theory of Markov processes. The trajectory is built in a randomized manner: The successor of the current point is chosen stochastically, with a probability that depends only on the current point and not on the previous history.

$$Y \leftarrow \text{Neighbor}(N(X^{(t)}))$$

$$X^{(t+1)} \leftarrow \begin{cases} Y & \text{if } f(Y) \leq f(X^{(t)}) \\ Y & \text{if } f(Y) > f(X^{(t)}), \text{ with probability } p = e^{-(f(Y)-f(X^{(t)}))/T} \\ X^{(t)} & \text{if } f(Y) > f(X^{(t)}), \text{ with probability } (1-p). \end{cases} \quad (3.1)$$

SA introduces a *temperature* parameter T, which determines the probability that worsening moves are accepted: A larger T implies that more worsening moves tend to be accepted, and therefore a larger diversification occurs. The rule in (3.1) is called *exponential acceptance rule*. If T goes to infinity, then the probability that a move is accepted becomes 1, whether it improves the result or not, and one obtains a random walk. Vice versa, if T goes to zero, only improving moves are accepted as in the standard local search. Being a Markov process, SA is characterized by a memory-less property: If one starts the process and waits long enough, the memory about the initial configuration is lost and the probability of finding a given configuration at a given state will be stationary and only dependent on the value of f. If T goes to zero, the probability will peak only at the globally optimal configurations. This basic result raised high hopes of solving optimization problems through a simple and general-purpose method, starting from seminal work in physics [184] and in optimization [1, 63, 161, 211].

Unfortunately, after some decades, it became clear that SA is not a panacea. Furthermore, most mathematical results about asymptotic convergence (converge of the method when the number of iterations goes to infinity) are quite irrelevant for optimization. First, one does not care whether the *final* configuration at convergence is optimal or not, but that an optimal solution (or a good approximation thereof) is encountered – and memorized – during the search. Second, asymptotic convergence usually requires a patience that is excessive considering the limited length of our lives. Actually, repeated local search [95], and even random search [64], have better asymptotic results for some problems.

3.2 Simulated Annealing and Asymptotics

Given the limited practical relevance of the theoretical results, a practitioner has to place asymptotic results in the background and develop heuristics where high importance is attributed to learning from a task before the search is started and from the current local characteristics during the search. In the following, after a brief introduction about some asymptotic convergence results, we will concentrate on the more recent developments in SA considering adaptive and learning implementations.

3.2.1 Asymptotic Convergence Results

Let $(\mathscr{X}, f)$ be an instance of a combinatorial optimization problem, $\mathscr{X}$ being the search space and f being the objective function. Let $\mathscr{X}^*$ be the set of optimal solutions. One starts from an initial configuration $X^{(0)}$ and repeatedly applies (3.1) to generate a trajectory $X^{(t)}$. Under appropriate conditions, the probability of finding one of the optimal solutions tends to one when the number of iterations goes to infinity:

$$\lim_{k \to \infty} \Pr(X^{(k)} \in \mathscr{X}^*) = 1. \tag{3.2}$$

Let $\mathscr{O}$ denote the set of possible outcomes (states) of a sampling process, let $X^{(k)}$ be the stochastic variable denoting the outcome of the kth trial, then the elements of the *transition probability matrix* P, given the probability that the configuration is at a specific state j given that it was at state i before the last step, are defined as:

$$p_{ij}(k) = \Pr(X^{(k)} = j | X^{(k-1)} = i). \tag{3.3}$$

A *stationary distribution* of a finite time-*homogeneous* (meaning that transitions do not depend on time) Markov chain is defined as the stochastic vector q whose components are given by

$$q_i = \lim_{k \to \infty} \Pr(X^{(k)} = i | X^{(0)} = j) \quad \text{for all } j \in \mathscr{O}. \tag{3.4}$$

If a stationary distribution exists, one has $\lim_{k \to \infty} \Pr(X^{(k)} = i) = q_i$. Furthermore $q^{\mathrm{T}} = q^{\mathrm{T}} P$, the distribution is not modified by a single Markov step.

If a finite Markov chain is homogeneous, *irreducible* (for every i, j, there is a positive probability of reaching i from j in a finite number of steps) and *aperiodic* (the greatest common divisor $\gcd(\mathscr{D}_i) = 1$, where $\mathscr{D}_i$ is the set of all integers $n > 0$ with $(P^n)_{ii} > 0$), there exist a unique stationary distribution, determined by the equation:

$$\sum_{j \in \mathscr{O}} q_j p_{ji} = q_i. \tag{3.5}$$

3.2.1.1 Homogeneous Model

In the homogeneous model, one considers a sequence of infinitely long homogeneous Markov chains, where each chain is for a fixed value of the temperature T.

Under appropriate conditions [2] (the generation probability must ensure that one can move from an arbitrary initial solution to a second arbitrary solution in a finite number of steps), the Markov chain associated to SA has a stationary distribution $q(T)$ whose components are given by:

$$q_i(T) = \frac{e^{-f(i)/T}}{\sum_{j \in \mathscr{X}} e^{-f(j)/T}} \tag{3.6}$$

and

$$\lim_{T \to 0} q_i(T) = q_i^* = \frac{1}{|\mathscr{X}^*|} I_{\mathscr{X}^*}(i), \tag{3.7}$$

where $I_{\mathscr{X}^*}$ is the characteristic function of the set $\mathscr{X}^*$, equal to one if the argument belongs to the set, zero otherwise.

It follows that:

$$\lim_{T \to 0} \lim_{k \to \infty} \Pr_T(X^{(k)} \in \mathscr{X}^*) = 1. \tag{3.8}$$

The algorithm asymptotically finds an optimal solution with probability one, "converges with probability one."

3.2.1.2 Inhomogeneous Model

In practice, one cannot wait for a stationary distribution to be reached. The temperature must be lowered before converging. At each iteration k one therefore has a different temperature T_k, obtaining a nonincreasing sequence of values T_k such that $\lim_{k \to \infty} T_k = 0$.

If the temperature decreases in a sufficiently slow way:

$$T_k \geq \frac{A}{\log(k+k_0)} \tag{3.9}$$

for $A > 0$ and $k_0 > 2$; then the Markov chain converges in distribution to q^* or, in other words

$$\lim_{k \to \infty} \Pr(X^{(k)} \in \mathscr{X}^*) = 1. \tag{3.10}$$

The theoretical value of A depends on the depth of the deepest local, nonglobal optimum, a value that is not easy to calculate for a generic instance.

The asymptotic convergence results of SA in both the homogeneous and inhomogeneous model are unfortunately *irrelevant for the application of SA to optimization*. In any finite-time approximation one must resort to approximations of the asymptotic convergence. The "speed of convergence" to the stationary distribution is determined by the second largest eigenvalue of the transition probability matrix $P(T)$ (not easy to calculate!). The number of transitions is at least *quadratic* in the total number of possible configurations in the search space [2]. For the inhomogeneous case, it can happen (e.g., Traveling Salesman Problem) that the *complete enumeration of all solutions* would take less time than approximating an optimal solution arbitrarily closely by SA [2].

In addition, repeated local search [95], and even random search [64], has better asymptotic results. According to [2] "approximating the asymptotic behavior of SA arbitrarily closely requires a number of transitions that for most problems is typically larger than the size of the solution space. Thus, the SA algorithm is clearly unsuited for solving combinatorial optimization problems to optimality." Of course, SA can be used in practice with fast *cooling schedules*, i.e., ways to progressively reduce the temperature during the search, but then the asymptotic results are not directly applicable. The optimal finite-length annealing schedules obtained on specific simple problems do not always correspond to those expected from the limiting theorems [244].

More details about cooling schedules can be found in [120, 140, 187]. Extensive experimental results of SA for graph partitioning, coloring, and number partitioning are presented in [151, 152]. A comparison of SA and reactive search is presented in [34, 35].

3.3 Online Learning Strategies in Simulated Annealing

If one wants to be poetic, the main feature of simulated annealing lies in its asymptotic convergence properties; the main drawback lies in the *asymptotic* convergence. For a practical application of SA, if the local configuration is close to a local minimizer and the temperature is already very small in comparison to the upward jump that has to be executed to escape from the attractor, although the system will *eventually* escape, an enormous number of iterations can be spent around the attractor. Given a finite patience time, all future iterations can be spent while "circling like a fly around a light-bulb" (the light-bulb being a local minimum). Animals with superior cognitive abilities get burnt once, learn, and avoid doing it again!

The memoryless property (current move depending only on the current state, not on the previous history) makes SA look like a dumb animal indeed. It is intuitive that a better performance can be obtained by using memory, by self-analyzing the evolution of the search, by developing simple models, and by activating more direct *escape* strategies aiming at a better time-management than the "let's go to infinite time" principle. In the following sections, we summarize the main memory-based approaches developed in the years to make SA a competitive strategy.

3.3.1 Combinatorial Optimization Problems

Even if a vanilla version of a cooling schedule for SA is adopted (starting temperature T_{start}, geometric cooling schedule $T_{t+1} = \alpha\, T_t$, with $\alpha < 1$, final temperature T_{end}), a sensible choice has to be made for the three involved parameters T_{start}, α, and T_{end}. If the scale of the temperature is wrong, extremely poor results are to be expected. The work [266] suggests to estimate the distribution of f values; note that f is usually called "energy" when exploiting the physical analogies of SA. The standard deviation of the energy distribution defines the maximum-temperature scale, while the minimum change in energy defines the minimum-temperature scale. These temperature scales tell us where to begin and end an annealing schedule.

The analogy with physics is pursued in [166], where concepts related to *phase transitions* and *specific heat* are used. While we avoid entering physics, the idea is that a phase transition is related to solving a subpart of a problem. Before reaching the state after the transition, big reconfigurations take place and this is signaled by wide variations of the f values. In detail, a phase transition occurs when the specific heat is maximal, a quantity estimated by the ratio between the estimated variance of the objective function and the temperature: σ_f^2/T. After a phase transition corresponding to the big reconfiguration occurs, finer details in the solution have to be fixed, and this requires heuristically a slower decrease of the temperature. Concretely, one defines two temperature-reduction parameters α and β, monitors the evolution of f along the trajectory, and after the phase transition takes place at a given T_{msp}, one switches from a faster temperature decrease given by α to the slower one given by β. The value T_{msp} is the temperature corresponding to the maximum specific heat, when the scaled variance reaches its maximal value.

Up to now we discussed only about a monotonic decrease of the temperature. This process has some weaknesses: For fixed values of T_{start} and α in the vanilla version, one will reach an iteration so that the temperature will be so low that *practically* no tentative move will be accepted with a nonnegligible probability (given the finite users' patience). The best value reached so far, f_{best}, will remain stuck in a helpless manner even if the search is continued for very long CPU times, see also Fig. 3.1. In other words, given a set of parameters T_{start} and α, the useful span of CPU time is practically limited. After the initial period, the temperature will be so low that the system *freezes* and, with large probability, no tentative moves will be accepted anymore within the finite span of the run. Now, for a new instance, it is not so simple to guess appropriate parameter values. Furthermore, in many cases one would like to use an *anytime algorithm*, so that longer allocated CPU times are related to possibly better and better values until the user decides to stop. Anytime algorithms – by definition – return the best answer possible even if they are not allowed to run to completion, and may improve on the answer if they are allowed to run longer.

Let us note that, in many cases, the stopping criterion should be decided a posteriori, for example, after determining that additional time has little probability to improve significantly on the result.

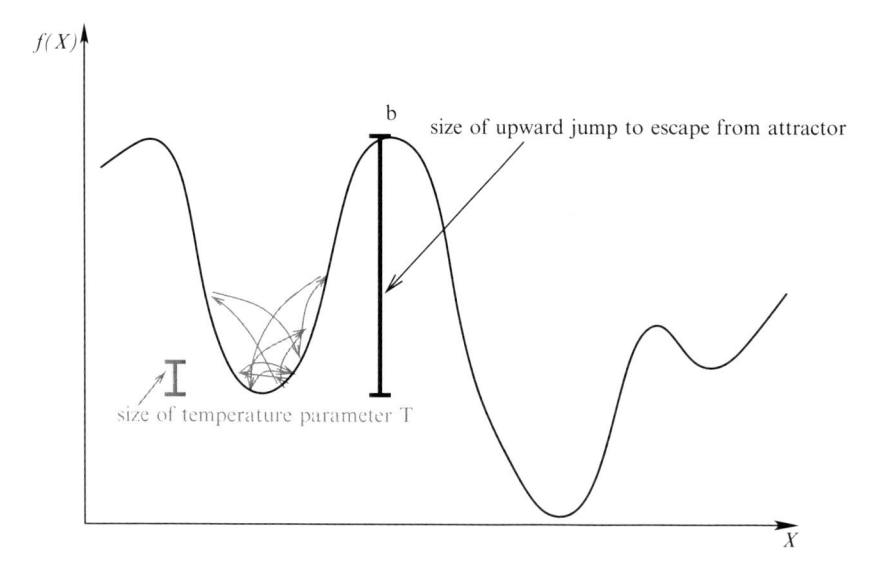

3.3 Online Learning Strategies in Simulated Annealing

Fig. 3.1 Simulated annealing: If the temperature is very low with respect to the jump size, SA risks a practical entrapment close to a local minimizer

Because this problem is related to a monotonic temperature decrease, a motivation arises to consider *nonmonotonic cooling schedules*; see [3, 73, 201]. A very simple proposal [73] suggests to reset the temperature once and for all at a constant temperature high enough to escape local minima but also low enough to visit them, for example, at the temperature T_{found} when the best heuristic solution is found in a preliminary SA simulation.

The basic design principle for a nonmonotonic schedule is related to (i) exploiting an attraction basin rapidly by decreasing the temperature so that the system can settle down close to the local minimizer, (ii) *increasing the temperature* to diversify the solution and visit other attraction basins, and (iii) decreasing again after reaching a different basin. As usual, the temperature increase in this kind of nonmonotonic cooling schedule has to be rapid enough to avoid falling back to the current local minimizer, but not too rapid to avoid a random-walk situation (where all random moves are accepted), which would not capitalize on the local structure of the problem ("good local minima close to other good local minima"). The implementation details have to do with determining an *entrapment* situation, for example, from the fact that no tentative move is accepted after a sequence $t_{\max}$ of tentative changes, and determining the detailed temperature decrease–increase evolution as a function of events occurring during the search. Possibilities to increase the temperature include resetting the temperature to $T_{\text{reset}} = T_{\text{found}}$, the temperature value when the current best solution was found [201]. If the reset is successful, one may progressively reduce the reset temperature: $T_{\text{reset}} \leftarrow T_{\text{reset}}/2$. Alternatively [3] geometric *reheating* phases can be used, which multiply T by a heating factor γ larger than one at each iteration during reheat. Enhanced versions involve a learning process to choose a proper value of the heating factor depending on the system state. In particular, γ is close to one at the beginning, while it increases if, after a fixed number of escape trials, the system is still trapped in the local minimum. More details and additional bibliography can be found in the cited papers.

Let us note that similar "strategic oscillations" have been proposed in tabu search, in particular in the reactive tabu search [33], see Chap. 4, and in variable neighborhood search, see Chap. 2.

Modifications departing from the exponential acceptance rule and adaptive stochastic local search methods for combinatorial optimization are considered in [194, 195]. Experimental evidence shows that the a priori determination of SA parameters and acceptance function does not lead to efficient implementations. Adaptations may be done *"by the algorithm itself using some learning mechanism* or by the user using his own learning mechanism." The authors appropriately note that the optimal choices of algorithm parameters depend not only on the problem but also on the particular instance and that a proof of convergence to a globally optimum is not a selling point for a specific heuristic: In fact a simple random sampling, or even exhaustive enumeration (if the set of configurations is finite), will eventually find the optimal solution, although they are not the best algorithms to suggest. A simple adaptive technique suggested in [195] is the SEQUENCE-HEURISTIC: A perturbation leading to a worsening solution is accepted if and only if a fixed number of trials could not find an improving perturbation. This method can be seen as deriving evidence of "entrapment" in a local minimum and reactively activating an escape mechanism. In this way the temperature parameter is eliminated. The positive performance of the SEQUENCEHEURISTIC in the area of design automation suggests that the success of SA is "due largely to its acceptance of bad perturbations to escape from local minima rather than to some mystical connection between combinatorial problems and the annealing of metals" [195].

"Cybernetic" optimization is proposed in [96] as a way to use probabilistic information for feedback during a run of SA. The idea is to consider more runs of SA running in parallel and to aim at *intensifying the search* (by lowering the temperature parameter) when there is evidence that the search is converging to the optimum value. If one looks for gold, one should spend more time "looking where all the other prospectors are looking" [96] (but let us note that actually one may argue differently depending on the luck of the other prospectors!). The empirical evidence is taken from the similarity between current configurations of different parallel runs. For sure, if more solutions are close to the same optimum point, they are also close to each other. The contrary is not necessarily true; nonetheless, this similarity is taken as evidence of a (possible) closeness to the optimum point, implying intensification and causing in a reactive manner a gradual reduction of the temperature.

3.3.2 Global Optimization of Continuous Functions

The application of SA to continuous optimization (optimization of functions defined on real variables in $\mathbb{R}$) is pioneered by [75]. The basic method is to generate a new point with a random step along a direction e_h, to evaluate the function and to accept the move with the probability given in (3.1). One cycles over the different directions e_h during successive steps of the algorithm. A first critical choice has to do with the range of the random step along the chosen direction. A fixed choice obviously may be very inefficient: This opens a first possibility for *learning* from the local f surface. In particular, a new trial point x' is obtained from the current point x as:

$$x' = x + \text{RAND}(-1,1)v_h e_h,$$

where $\text{RAND}(-1,1)$ returns a random number uniformly distributed between -1 and 1, e_h is the unit-length vector along direction h, and v_h is the step-range parameter, one for each dimension h, collected into the vector v. The exponential acceptance rule is used to decide whether to update the current point with the new point x'. The v_h value is adapted during the search with the aim of maintaining the number of *accepted* moves at about one-half of the total number of tried moves. In particular, after a number N_S of random moves cycling along the coordinate directions, the step-range vector v is updated: Each component is increased if the number of accepted moves is greater than 60%, reduced if it is less than 40%. The speed of increase or decrease could be different for the different coordinate dimensions (in practice it is fixed to 2 and 1/2 in the above paper). After $N_S N_T$ cycles, (N_T being a second fixed parameter), the temperature is reduced in a multiplicative manner: $T_{k+1} \leftarrow r_T T_k$, and the current point is reset to the best-so-far point found during the previous search. Although the implementation is already reactive and based on memory, the authors encourage more work so that a "good monitoring of the minimization process" can deliver precious feedback about some crucial internal parameters of the algorithm.

In adaptive simulated annealing (ASA), also known as very fast simulated re-annealing [144], the parameters that control the temperature cooling schedule and the random step selection are automatically adjusted according to algorithm progress. If the state is represented as a point in a box and the moves as an oval cloud around it, the temperature and the step size are adjusted so that all of the search space is sampled at a coarse resolution in the early stages, while the state is directed to promising areas in the later stages [144].

Chapter 4
Reactive Prohibitions

> *It is a good morning exercise for a research scientist to discard a pet hypothesis every day before breakfast. It keeps him young.*
> *(Konrad Lorenz)*

4.1 Prohibitions for Diversification

The picturesque example about bicycle design in Fig. 2.1 was intended to motivate the fact that *prohibiting* some moves, for example, moves canceling recent local changes, and persisting stubbornly in spite of seemingly negative results can have positive effects to unblock a suboptimal configuration and eventually to identify better solutions to complex problems. Similarly, the above citation by Lorenz is reminding us that scientific activity is often related to "discarding pet hypotheses," i.e., to prohibiting the use of some previously available solutions in order to catalyze real innovation.

The tabu search (TS) metaheuristic [107] is based on the use of *prohibition-based* techniques and "intelligent" schemes as a complement to basic heuristic algorithms such as local search, with the purpose of guiding the basic heuristic *beyond local optimality*. It is difficult to assign a precise date of birth to these principles. For example, ideas similar to those proposed in TS can be found in the *denial* strategy of [241] (once common features are detected in many suboptimal solutions, they are *forbidden*) or in the opposite *reduction* strategy of [172] (in an application to the Traveling Salesman Problem, all edges that are common to a set of local optima are fixed). In very different contexts, prohibition-like strategies can be found in *cutting planes* algorithms for solving integer problems through their linear programming relaxation (inequalities that cut off previously obtained fractional solutions are generated) and in branch and bound algorithms (subtrees are not considered if their leaves cannot correspond to better solutions); see the textbook [205].

The renaissance and full blossoming of "intelligent prohibition-based heuristics" starting from the late 1980s is greatly because of the role of Glover in the proposal and diffusion of a rich variety of metaheuristic tools [107, 108], but see also [124] for an independent seminal paper. It is evident that Glover's ideas have been a source of inspiration for many approaches based on the intelligent use of memory in heuristics. Let us only cite the four "principles" for using long-term memory in tabu-search of *recency, frequency, quality, and influence*, also cited in [45]. A growing

Table 4.1 A classification of TS methods based on discrete dynamical systems

	Discrete Dynamical System (search trajectory generation)	
	Deterministic	Stochastic
	Strict-TS	Probabilistic-TS
	Fixed-TS	Robust-TS
		Fixed-TS with stochastic tie breaking
	Reactive-TS	Reactive-TS with stochastic tie breaking
		Reactive-TS with neighborhood sampling
		(stochastic candidate list strategies)

number of TS-based algorithms has been developed in the last years and applied with success to a wide selection of problems [109]. It is therefore difficult, if not impossible, to characterize a "canonical form" of TS, and classifications tend to be short-lived. Nonetheless, at least two aspects characterize many versions of TS: the fact that TS is used to complement *local (neighborhood) search*, and the fact that the main modifications to local search are obtained through the *prohibition* of selected moves available at the current point. TS acts to continue the search beyond the first local minimizer without wasting the work already executed, and to enforce appropriate amounts of diversification to avoid that the search trajectory remains confined near a given local minimizer.

In our opinion, the main competitive advantage of TS with respect to alternative heuristics based on local search such as simulated annealing (SA) [161] lies in the intelligent use of the past history of the search to influence its future steps. Because TS includes now a wide variety of methods, we use the term *prohibition-based search* when the investigation is focused onto the use of prohibition to encourage diversification. In any case, the term TS is used with the same meaning in this chapter.

Let us assume that the feasible search space is the set of binary strings with a given length L: $\mathscr{X} = \{0,1\}^L$. $X^{(t)}$ is the current configuration and $N(X^{(t)})$ the previously introduced neighborhood. In prohibition-based search, some of the neighbors are *prohibited*, a subset $N_A(X^{(t)}) \subset N(X^{(t)})$ contains the *allowed* ones. The general way of generating the search trajectory that we consider is given by:

$$X^{(t+1)} = \text{Best-Neighbor}(N_A(X^{(t)})) \qquad (4.1)$$
$$N_A(X^{(t+1)}) = \text{Allow}(N(X^{(t+1)}), X^{(0)}, ..., X^{(t+1)}). \qquad (4.2)$$

The set-valued function ALLOW selects a subset of $N(X^{(t+1)})$ in a manner that depends on the entire search trajectory $X^{(0)}, ..., X^{(t+1)}$.

4.1.1 Forms of Prohibition-Based Search

There are several points of views to classify heuristic algorithms. By analogy with the concept of *abstract data type* in computer science [4], and with the related *object-oriented* software engineering framework [77], it is useful to separate the abstract concepts and operations of TS from the detailed implementation, i.e., realization with specific data structures. In other words, *policies* (that determine which trajectory is generated in the search space, what the balance of intensification and diversification is, etc.) should be separated from *mechanisms* that determine *how* a specific policy is realized. An essential abstract concept in TS is given by the *discrete dynamical system* of (4.1)–(4.2) obtained by modifying local search.

4.1.2 Dynamical Systems

A classification of some prohibition-based algorithms that is based on the underlying dynamical system is illustrated in Table 4.1.

A first subdivision is given by the *deterministic* vs. *stochastic* nature of the system. Let us first consider the deterministic versions. Possibly the simplest form of TS is what is called *strict-TS*: A neighbor is prohibited if and only if it has already been visited during the previous part of the search [107] (the term strict is chosen to underline the rigid enforcement of its simple prohibition rule). Therefore, (4.2) becomes

$$N_A(X^{(t+1)}) = \{X \in N(X^{(t+1)}) \quad \text{s. t.} \quad X \notin \{X^{(0)},...,X^{(t+1)}\}\}. \quad (4.3)$$

Let us note that strict-TS is parameter-free.

Two additional algorithms can be obtained by introducing a *prohibition parameter* T that determines how long a move will remain prohibited after the execution of its inverse. The *fixed-TS* algorithm is obtained by fixing T throughout the

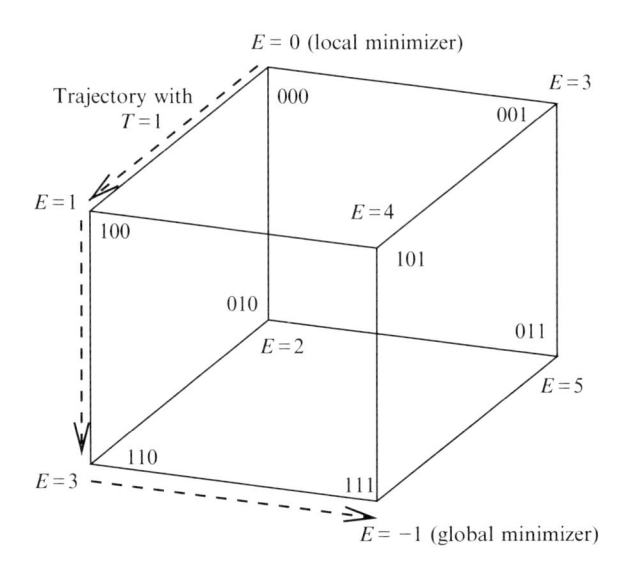

Fig. 4.1 Search space, f values, and search trajectory

search [107]. Let μ^{-1} denote the *inverse* of a move, for example, if μ_i is changing the ith bit of a binary string from 0 to 1, μ_i^{-1} changes the same bit from 1 to 0. A neighbor is allowed if and only if it is obtained from the current point by applying a move such that its inverse has not been used during the last T iterations. In detail, if LASTUSED(μ) is the last usage time of move μ (LASTUSED(μ) = $-\infty$ at the beginning):

$$N_A(X^{(t)}) = \{X = \mu \circ X^{(t)} \quad \text{s. t. LASTUSED}(\mu^{-1}) < (t-T)\}. \qquad (4.4)$$

If T changes with the iteration counter depending on the search status, and in this case the notation is $T^{(t)}$, the general dynamical system that generates the search trajectory comprises an additional evolution equation for $T^{(t)}$, so that the three defining equations are now:

$$T^{(t)} = \text{REACT}(T^{(t-1)}, X^{(0)}, ..., X^{(t)}) \qquad (4.5)$$
$$N_A(X^{(t)}) = \{X = \mu \circ X^{(t)} \quad \text{s. t. LASTUSED}(\mu^{-1}) < (t - T^{(t)})\} \qquad (4.6)$$
$$X^{(t+1)} = \text{BEST-NEIGHBOR}(N_A(X^{(t)})). \qquad (4.7)$$

Let us note that the prohibition rule has a very simple implementation for basic moves acting on binary strings. The prohibition rule is in this case: after changing a bit, "do not touch the same bit again for the next T iterations." Think about enclosing the just changed bit in an ice cube that will melt down only after T iterations. Now, possible rules to determine the prohibition parameter by reacting to the repetition of previously-visited configurations have been proposed in [33] (*reactive-TS, RTS*). In addition, there are situations where the single reactive mechanism on T is not sufficient to avoid long cycles in the search trajectory and therefore a second reaction is needed [33]. The main principles of RTS are briefly reviewed in Sect. 4.2.

Stochastic algorithms related to the previously described deterministic versions can be obtained in many ways. For example, prohibition rules can be substituted with *probabilistic generation-acceptance rules* with large probability for allowed moves, small for prohibited ones, see for example the *probabilistic-TS* [107]. Stochasticity can increase the robustness of the different algorithms. Citing from [107], "randomization is a means for achieving diversity without reliance on memory," although it could "entail a loss in efficiency by allowing duplications and potentially unproductive wandering that a more systematic approach would seek to eliminate." Incidentally, asymptotic results for TS can be obtained in probabilistic-TS [93]. In a different proposal (*robust-TS*) the prohibition parameter is randomly changed between an upper and a lower bound during the search [248]. Stochasticity in fixed-TS and in reactive-TS can be added through a *random breaking of ties*, in the event that the same cost function decrease is obtained by more than one winner in the BEST-NEIGHBOR computation. At least this simple form of stochasticity should always be used to avoid external search biases, possibly caused by the ordering of the loop indices.

4.1 Prohibitions for Diversification

If the neighborhood evaluation is expensive, the exhaustive evaluation can be substituted with a partial *stochastic sampling*: Only a partial list of candidates is examined before choosing the best allowed neighbor.

Finally, other possibilities that are softer than prohibitions exist. For example, the HSAT [106] variation of GSAT introduces a tie-breaking rule into GSAT: If more moves produce the same (best) Δf, the preferred move is the one that has not been applied for the longest span. HSAT can be seen as a "soft" version of TS: while TS prohibits recently-applied moves, HSAT discourages recent moves if the same Δf can be obtained with moves that have been "inactive" for a longer time. It is remarkable to see how this innocent variation of GSAT can increase its performance on some SAT benchmark tasks [106].

4.1.3 A Worked-Out Example of Fixed Tabu Search

Let us assume that the search space $\mathscr{X}$ is the set of 3-bit strings ($\mathscr{X} = [b_1, b_2, b_3]$) and the cost function is:

$$f([b_1,b_2,b_3]) = b_1 + 2\,b_2 + 3\,b_3 - 7\,b_1\,b_2\,b_3.$$

The feasible points (the edges of the three-dimensional binary cube) are illustrated in Fig. 4.1 with the associated cost function. The neighborhood of a point is the set of points that are connected with edges.

The point $X^{(0)} = [0,0,0]$ with $f(X^{(0)}) = 0$ is a local minimizer because all moves produce a higher cost value. The best of the three admissible moves is μ_1, so that $X^{(1)} = [1,0,0]$. Note that the move is applied even if $f(X^{(1)}) = 1 \geq f(X^{(0)})$, so that the system abandons the local minimizer.

If $T^{(1)} = 0$, the best move from $X^{(1)}$ will again be μ_1 and the system will return to its starting point: $X^{(1)} = X^{(0)}$. If $T^{(t)}$ remains equal to zero, the system is trapped forever in the limit cycle $[0,0,0] \to [1,0,0] \to [0,0,0] \to [1,0,0]\ldots$.

On the contrary, if $T^{(t)} = 1$, μ_1 is prohibited at $t = 1$ because it was used too recently, i.e. its most recent usage time $\text{LASTUSED}(\mu_1)$ satisfies $\text{LASTUSED}(\mu_1) = 0 \geq (t - T^{(t)}) = 0$. The neighborhood is therefore limited to the points that can be reached by applying μ_2 or μ_3 ($N([1,0,0]) = \{[1,1,0],[1,0,1]\}$). The best *admissible* move is μ_2, so that $X^{(2)} = [1,1,0]$ with $f(X^{(2)}) = 3$.

At $t = 2$ μ_2 is prohibited, μ_1 is admissible again because $\text{LASTUSED}(\mu_1) = 0 < (t - T^{(t)}) = 1$, and μ_3 is admissible because it was never used. The best move is μ_3 and the system reaches the global minimizer: $X^{(3)} = [1,1,1]$ with $f(X^{(3)}) = -1$.

4.1.4 Relationship Between Prohibition and Diversification

The prohibition parameter T used in (4.4) is related to the amount of *diversification*: the larger T, the longer the distance that the search trajectory must go before it is allowed to come back to a previously visited point. In particular, the following relationships between prohibition and diversification are demonstrated in [23] for a search space consisting of binary strings with basic moves flipping individual bits:

- The Hamming distance H between a starting point and successive point along the trajectory is strictly increasing for $T+1$ steps.

$$H(X^{(t+\Delta t)}, X^{(t)}) = \Delta t \quad \text{for} \quad \Delta t \leq T+1.$$

- The minimum repetition interval R along the trajectory is $2(T+1)$.

$$X^{(t+R)} = X^{(t)} \Rightarrow R \geq 2(T+1).$$

The linear increase of the distance is caused by the changed bits that are frozen for T iterations. Each changed bit increments the distance by one. As shown in Fig. 4.2, only when the first changed, and frozen, bit melts down (its prohibition expires) it can be changed back to its original value so that the Hamming distance can decrease.

The second proposition is demonstrated if one considers that, as soon as the value of a variable is flipped, this value remains unchanged for the next T iterations. To come back to the starting configuration, all $T+1$ variables changed in the first phase

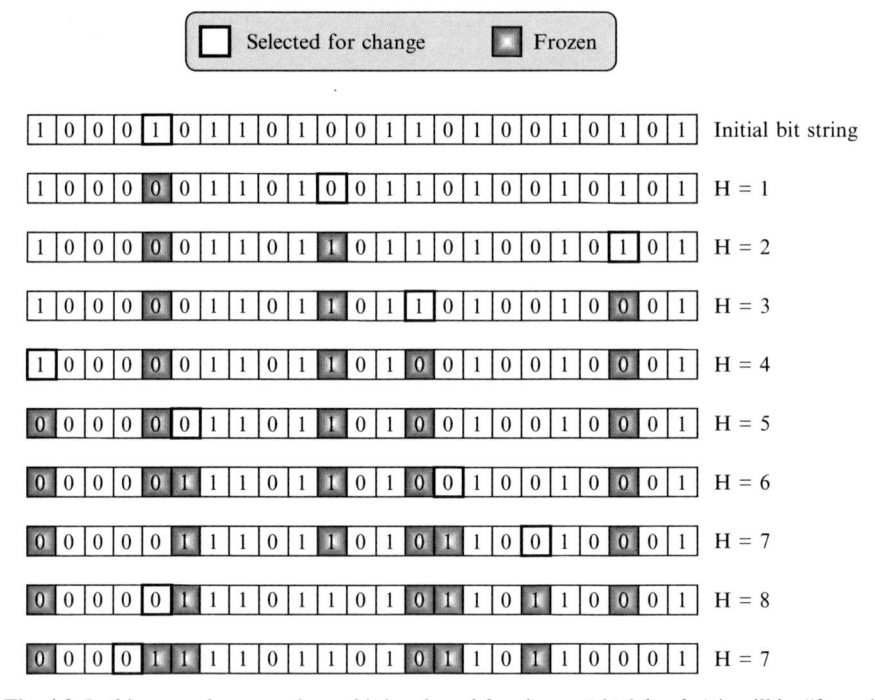

Fig. 4.2 In this example, every time a bit is selected for change (*thick border*) it will be "frozen" (*grey cells*), i.e., it will not be selected again for T steps, causing the Hamming distance to increase at least up to $(T+1)$. In the example, $T=6$

must be changed back to their original values. By the way, because of the prohibition scheme, if the trajectory comes back as soon as possible to a point visited in the past (for example, because the attraction basin around a local optimum is too large to be surpassed with the given T value), it does so by visiting different configurations with respect to those visited while leaving the point. The mental image is that of a loop (a lasso) in configuration space, not that of a linear oscillation caused by a spring connected to a local optimum.

But large values of T imply that only a limited subset of the possible moves can be applied to the current configuration. In particular, if L is the number of bits in the string, T must be less than or equal to $L-2$ to assure that at least two moves can be applied. Clearly, if only one move can be applied, the choice is not influenced by the f values in the neighborhood and one cannot hope to identify good quality solutions. It is therefore appropriate to set T to the smallest value that guarantees diversification.

4.1.5 How to Escape from an Attractor

Local minima points are *attractors* of the search trajectory generated by deterministic local search. If the cost function is integer-valued and lower bounded, it can be easily shown that a trajectory starting from an arbitrary point will terminate at a local minimizer. All points such that a deterministic local search trajectory starting from them terminates at a specific local minimizer make up its *attraction basin*. Figure 4.3 shows in its upper part an objective function with three local minima, represented by contour lines.

Now, as soon as a local minimizer is encountered, its entire attraction basin (the three colored islands of Fig. 4.3, upper part) is not of interest for the optimization procedure, in fact its points do not have smaller cost values. It is nonetheless true that better points could be close to the current basin, whose boundaries are not known. One of the problems that must be solved in heuristic techniques based on local search is how to continue the search beyond the local minimizer and how to *avoid the confinement* of the search trajectory. Confinements can happen because the trajectory tends to be biased toward points with low cost function values, and therefore also toward the just abandoned local minimizer. The fact that the search trajectory remains close to the minimizer for some iterations is clearly a desired effect in the hypothesis that better points are preferentially located in the neighborhood of good suboptimal point rather than among randomly extracted points.

Simple confinements can be *cycles* (endless repetition of a sequence of configurations during the search) or more complex trajectories with no clear periodicity but nonetheless such that only a limited portion of the search space is visited (they are analogous to *chaotic attractors* in dynamical systems).

An heuristic prescription is that the search point is kept close to a discovered local minimizer at the beginning, snooping about better attraction basins. If these are not discovered, the search should gradually progress to larger distances (therefore progressively enforcing longer-term diversification strategies).

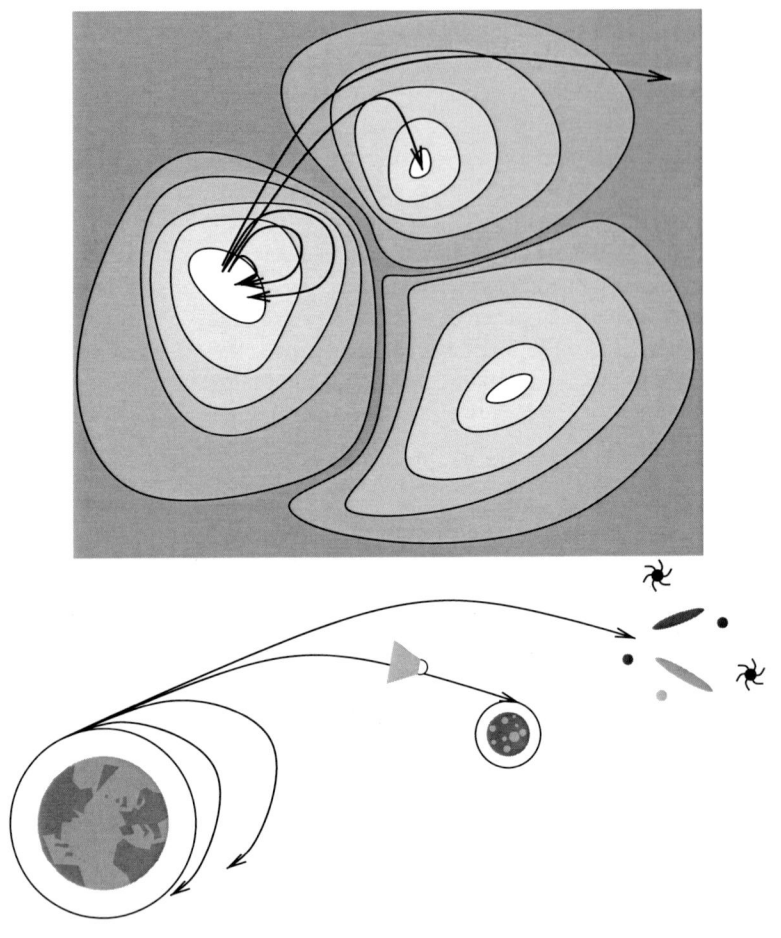

Fig. 4.3 Applying the minimum diversification needed to escape the local attractor. *Top*: If too little diversification is applied, the trajectory falls back to the same minimum; if diversification is excessive, nearby minima may not be visited. *Bottom*: the same concept applied to space travel: The right amount of diversification (plus ballistic calculations) allows the capsule to reach a suitable moon orbit; too little and the capsule falls back, too much and the target is missed

Some very different ways of realizing this general prescription are illustrated here for a "laboratory" test problem originally studied in [21]. The search space is the set of all binary strings of length L. Let us assume that the search has just reached a (strict) local minimizer and that the cost f in the neighborhood is strictly increasing as a function of the number of different bits with respect to the given local minimizer (i.e., as a function of the Hamming distance). Without loss of generality, let us assume that the local minimizer is the zero string ($[00...0]$) and that the cost is precisely the Hamming distance. Although artificial, the assumption is not unrealistic in many cases. An analogy in continuous space is the usual positive-definite

4.1 Prohibitions for Diversification

quadratic approximation of the cost in the neighborhood of a strict local minimizer of a differentiable function. In the following parts the discussion is mostly limited to deterministic versions.

4.1.5.1 Strict-TS

In the deterministic version of strict-TS, if more than one basic move produce the same cost decrease at a given iteration, the move that acts on the right-most (least significant) bit of the string is selected.

The set of obtained configuration for $L = 4$ is illustrated in Fig. 4.4. Let us now consider how the Hamming distance evolves in time, in the optimistic assumption that the search always finds an allowed move until all points of the search space are visited. If $H(t)$ is the Hamming distance at iteration t, the following holds true:

$$H(t) \leq \lfloor \log_2(t) \rfloor + 1. \tag{4.8}$$

This can be demonstrated after observing that a complete trajectory for an $(L-1)$-bit search space becomes a legal *initial part* of the trajectory for L-bit strings after appending zero as the most significant bit (see the trajectory for $L = 3$ in Fig. 4.4). Now, a certain Hamming distance H can be reached only as soon as or after the Hth bit is set (e.g, $H = 4$ can be reached only at or after $t = 8$ because the fourth bit is set at this iteration). Equation (4.8) trivially follows.

In practice the above optimistic assumption is not true: strict-TS can be stuck (trapped) at a configuration such that all neighbors have already been visited. In fact, the smallest L such that this event happens is $L = 4$ and the search is stuck at $t = 14$,

```
t = 0   H = 0  string: 0 0 0 0
t = 1   H = 1  string: 0 0 0 1
t = 2   H = 2  string: 0 0 1 1
t = 3   H = 1  string: 0 0 1 0
t = 4   H = 2  string: 0 1 1 0
t = 5   H = 1  string: 0 1 0 0
t = 6   H = 2  string: 0 1 0 1
t = 7   H = 3  string: 0 1 1 1
t = 8   H = 4  string: 1 1 1 1
t = 9   H = 3  string: 1 1 1 0
t = 10  H = 2  string: 1 1 0 0
t = 11  H = 1  string: 1 0 0 0
t = 12  H = 2  string: 1 0 0 1
t = 13  H = 3  string: 1 0 1 1
t = 14  H = 4  string: 1 0 1 0
              Stuck at t = 14
         (String not visited: 1101)
```

Trajectory for $L = 2$ (rows $t=0$ to $t=1$)
Trajectory for $L = 3$ (rows $t=0$ to $t=7$)

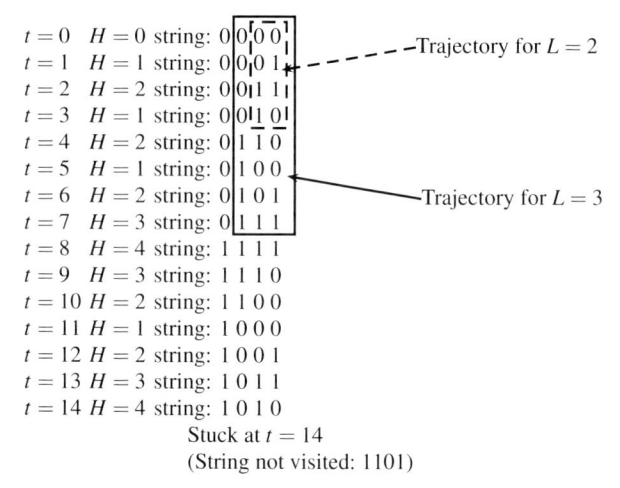

Fig. 4.4 Search trajectory for deterministic strict-TS: iteration t, Hamming distance H, and binary string

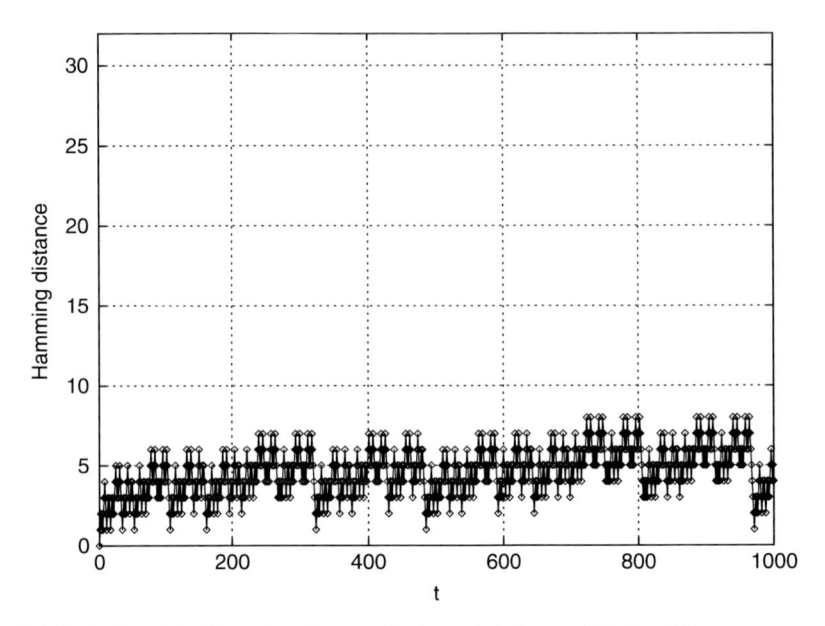

Fig. 4.5 Evolution of the Hamming distance for deterministic strict-TS ($L = 32$)

so that the string $[1101]$ is not visited. The problem worsens for higher-dimensional strings. For $L = 10$ the search is stuck after visiting 47% of the entire search space; for $L = 20$ it is stuck after visiting only 11% of the search space.

If the trajectory must reach Hamming distance H with respect to the local minimum point before escaping, i.e., before encountering a better attraction basin, the necessary number of iterations is at least exponential in H. Figure 4.5 shows the actual evolution of the Hamming distance for the case of $L = 32$. The detailed dynamics is complex, as "iron curtains" of visited points (that cannot be visited again) are created in the configuration space and the trajectory must obey the corresponding constraints. The slow growth of the Hamming distance is related to the "basin filling" effect [33] of strict-TS: All points at smaller distances tend to be visited before points at larger distances (unless the iron curtains prohibit the immediate visit of some points). On the other hand, the exploration is not as "intensifying" as an intuitive picture of strict-TS could lead one to believe: Some configurations at small Hamming distance are visited only in the last part of the search. As an example, the point at $H = 4$ is visited at $t = 8$ while one point at $H = 1$ is visited at $t = 11$ (t is the iteration). This is caused by the fact that, as soon as a new bit is set for the first time, all bits to the right are progressively cleared (because new configurations with lower cost are obtained). In particular, the second configuration at $H = 1$ is encountered at $t > 2$, the third at $t > 4$,... the n-th at $t > 2^{(n-1)}$. Therefore, at least a configuration at $H = 1$ will be encountered only after $2^{(L-1)}$ have been visited.

The relation of (4.8) is valid only in the assumption that strict-TS is deterministic and that it is not stuck for any configuration. Let us now assume that one manages to obtain a more "intensifying" version of strict-TS, i.e., that all configurations at

4.1 Prohibitions for Diversification

Hamming distance less than or equal to H are visited before configurations at distance greater than H. The initial growth of the Hamming distance in this case is much slower. In fact, the number of configurations C_H to be visited is:

$$C_H = \sum_{i=0}^{H} \binom{L}{i}. \tag{4.9}$$

It can be easily derived that $C_H \gg 2^H$, if $H \ll L$. As an example,[1] for $L = 32$ one obtains from (4.9) a value $C_5 = 242,825$, and therefore this number of configurations have to be visited before finding a configuration at Hamming distance greater than 5, while $2^5 = 32$. An explosion in the number of iterations spent near a local optimum occurs unless the nearest attraction basin is very close. The situation worsens in higher-dimensional search spaces: for $L = 64$, $C_5 = 8,303,633$, $C_4 = 679,121$. This effect can be seen as a manifestation of the "curse of dimensionality": A technique that works in very low-dimensional search space can encounter dramatic performance reductions as the dimension increases. In particular, there is the danger that the entire search span will be spent visiting points at small Hamming distances, unless additional diversifying tools are introduced.

4.1.5.2 Fixed-TS

The analysis of fixed-TS is simple: As soon as a bit is changed, it will remain prohibited ("frozen") for additional T steps. Therefore (see Fig. 4.6), the Hamming distance with respect to the starting configuration will cycle in a regular manner between zero and a maximal value $H = T + 1$ (only at this iteration the ice around the first frozen bit melts down and allows changes that are immediately executed because H decreases). All configurations in a cycle are different, apart from the initial configuration (see also the previous comments about "lasso" trajectories). The cycling T behavior is the same for both deterministic and stochastic versions. The property of the stochastic version is that different configurations have the possibility of being visited in different cycles. In fact all configurations at a given Hamming distance H have the same probability of being visited if $H \leq T + 1$, zero probability otherwise.

The effectiveness of fixed-TS in escaping from the local attractor depends on the size of the value T with respect to the minimal distance such that a new attraction basin is encountered. In particular, if T is too small, the trajectory will never escape, but if T is too large an "over-constrained" trajectory will be generated.

4.1.5.3 Simplified Reactive-TS

The behavior of reactive-TS depends on the specific reactive scheme used. Given the previously illustrated relation between the prohibition T and the diversification,

---
[1] Computations have been executed by Mathematica©.

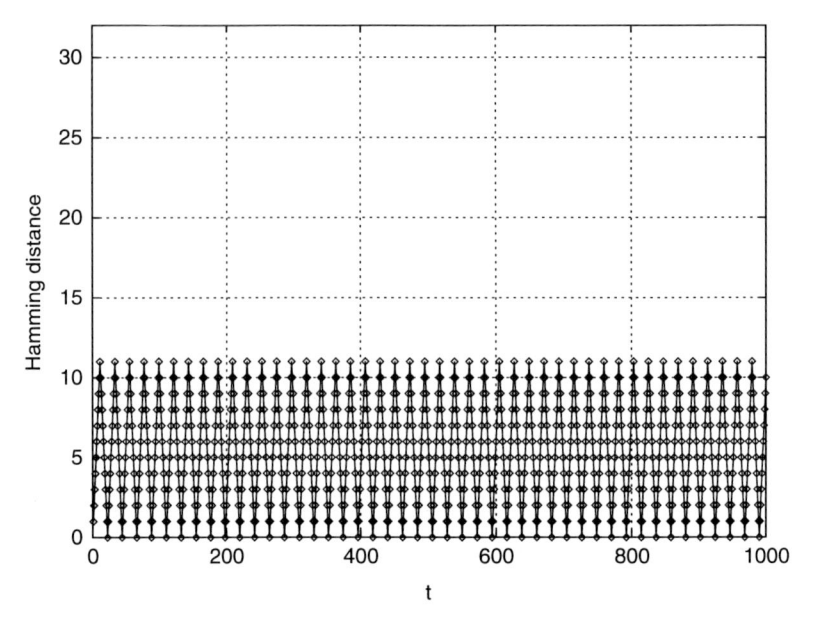

Fig. 4.6 Evolution of the Hamming distance for both deterministic and stochastic fixed-TS ($L = 32, T = 10$)

a possible prescription is that of gradually increasing T if there is evidence that the system is confined near a local attractor, until a new attractor is encountered. In particular, the evidence for a confinement can be obtained from the repetition of a previously visited configuration, while the fact that a new attraction basin has been found can be postulated if repetitions disappear for a suitably long period. In this last case, T is gradually decreased. This general procedure was used in the design of the RTS algorithm in [33], where specific rules are given for the entire feedback process.

In the present discussion, we consider only the initial phase of the escape from the attractor, when increases of T are dominant over decreases. In fact, to simplify the discussion, let us assume that $T = 1$ at the beginning and that the reaction acts to increase T when a local minimum point is repeated, in the following manner:

$$\text{REACT}(T) = \min\{\max\{T \times 1.1, T+1\}, L-2\}. \qquad (4.10)$$

For the prohibition parameter T, the initial value (and lower bound) of one implies that the system does not come back immediately to a just abandoned configuration. The upper bound is used to guarantee that not too many moves are prohibited and that at least two moves are allowed at each iteration. Nonintegral values of T are cast to integers before using them: They are cast to the largest integer less than or equal to T.

4.1 Prohibitions for Diversification

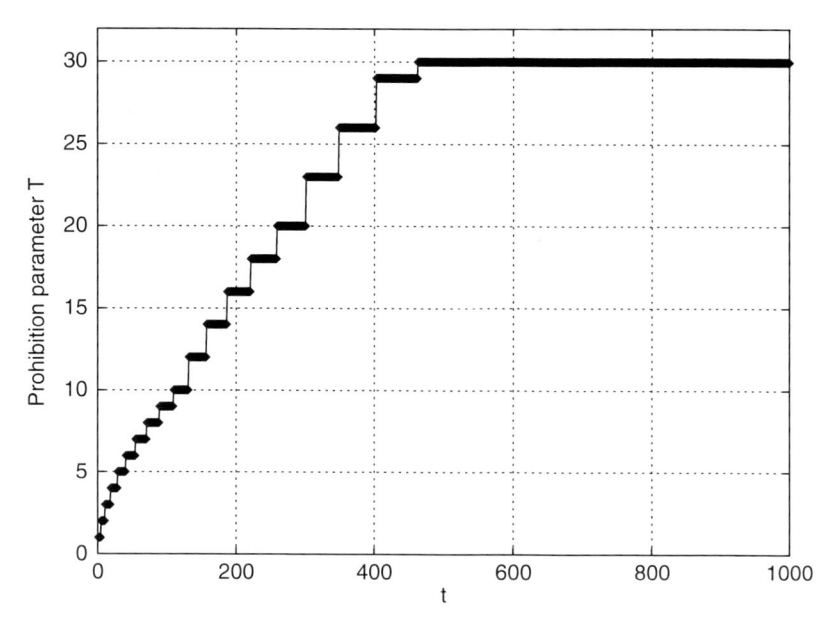

Fig. 4.7 Evolution of the prohibition parameter T for deterministic reactive-TS with reaction at local minimizers ($L = 32$)

The evolution of T for the deterministic version is shown in Fig. 4.7. Repeated occurrences of the local minimum point cause a rapid increase up to its maximal value. As soon as the maximum value is reached, the system enters a cycle. This limit cycle is caused by the fact that no additional attraction basins exist in the test case considered, while in real-world instances the prohibition T tends to be small with respect to its upper bound, both because of the limited size of the attraction basins and because of the complementary reaction that decreases T when repetitions disappear.

The behavior of the Hamming distance is illustrated in Fig. 4.8. The maximal Hamming distance reached increases in a much faster way compared with the strict-TS case.

Now, for a given $T^{(t)}$ the maximum Hamming distance that is reached during a cycle is $H_{\max} = T^{(t)} + 1$ and the cycle length is $2(T^{(t)} + 1)$. After the cycle is completed, the local minimizer is repeated and the reaction occurs. The result is that $T^{(t)}$ increases monotonically, and therefore the cycle length also does, as illustrated in Fig. 4.9 that expands the initial part of the graph.

Let us now consider a generic iteration t at which a reaction occurs (like $t = 4, 10, 18, \ldots$ in Fig. 4.9). At the beginning, (4.10) will increase T by one unit at each step. If, just before the reaction, the prohibition is T, the total number t of iterations executed is:

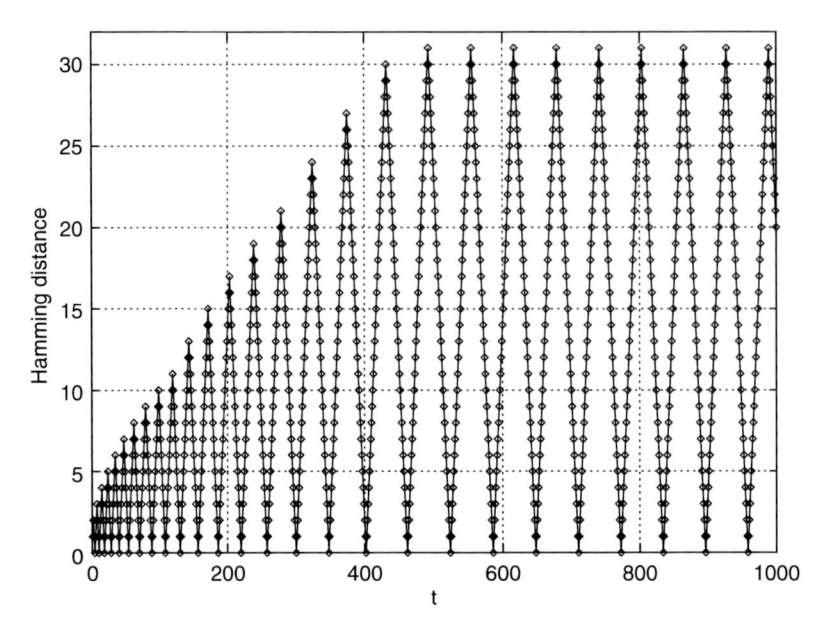

Fig. 4.8 Evolution of the Hamming distance for simplified reactive-TS ($L = 32$)

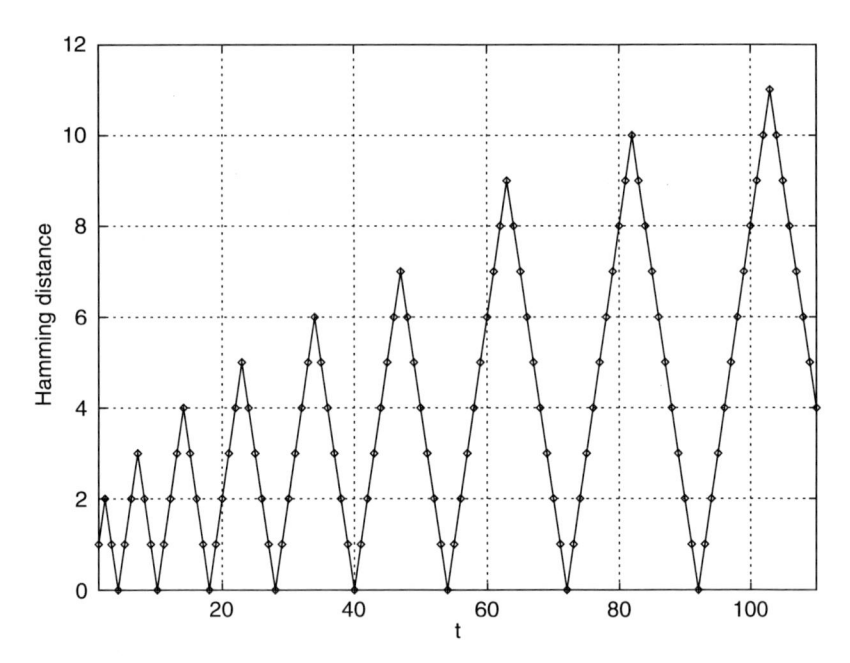

Fig. 4.9 Evolution of the Hamming distance for reactive-TS, first 100 iterations ($L = 32$)

$$t(T) = \sum_{i=1}^{T} 2(i+1) = 3T + T^2 \qquad (4.11)$$

$$t(H_{\max}) = \left(H_{\max}^2 + H_{\max} - 2\right) \qquad (4.12)$$

$$H_{\max}(t) = \frac{1}{2}\left(\sqrt{9+4t} - 1\right), \qquad (4.13)$$

where the relation $T = H_{\max} - 1$ has been used. Therefore, the increase of the maximum reachable Hamming distance is approximately $O(\sqrt{t})$ during the initial steps. The increase is clearly faster in later steps, when the reaction is multiplicative instead of additive (when $\lfloor T \times 1.1 \rfloor > T + 1$ in (4.10)), and therefore the above estimate of $H_{\max}$ becomes a lower bound in the following phase.

The difference with respect to strict-TS is a crucial one: One obtains an (optimistic) logarithmic increase in the strict algorithm, and a (pessimistic) increase that behaves like the square root of the number of iterations in the reactive case. In this last case bold tours at increasing distances are executed until the prohibition T is sufficient to escape from the attractor. In addition, if the properties of the fitness surface change slowly in the different regions, and RTS reaches a given local minimizer with a T value obtained during its previous history, the chances are large that it will escape even faster.

4.2 Reactive Tabu Search: Self-Adjusted Prohibition Period

A couple of issues arising in TS that are worth investigating are the determination of an appropriate prohibition T for the different tasks and different local configurations, in a way that is robust for a wide range of different problems, and the adoption of minimal computational complexity algorithms for using the search history. In particular, an *online* adjustment procedure for T is necessary if the instance being solved has different properties in different localities of its configuration space: The value of T needed to escape the left and right minima of the function in Fig. 4.10 (lightly shadowed regions) is larger than the ideal value for the central part, where a smaller distance is required to move from a local minimum to the next.

The issues are briefly discussed in the following sections, together with the RTS methods proposed to deal with them.

In RTS the prohibition T is determined through feedback (i.e., *reactive*) mechanisms during the search. T is equal to one at the beginning (the inverse of a given move is prohibited only at the next step), it increases only when there is *evidence* that diversification is needed, and it decreases when this evidence disappears. In detail: The evidence that diversification is needed is signaled by the repetition of previously visited configurations. All configurations found during the search are stored in memory. After a move is executed, the algorithm checks whether the current configuration has already been found and it reacts accordingly (T increases if a configuration is repeated, T decreases if no repetitions occurred during a sufficiently long period).

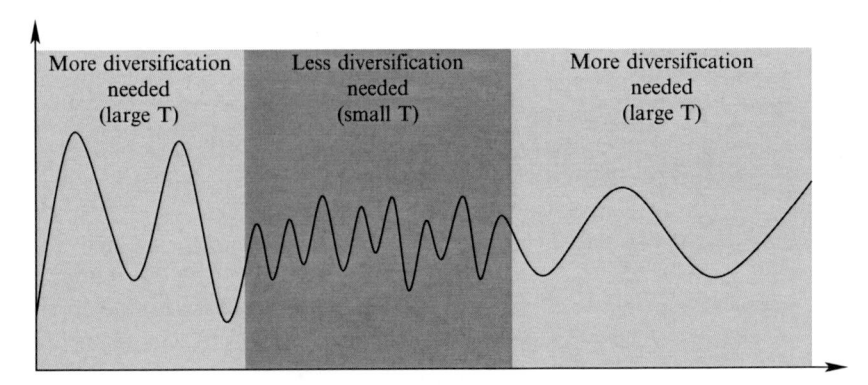

Fig. 4.10 The optimal value of the prohibition period T does not depend only on the problem or the instance: Different local landscapes may require Different values of T within the same run. For example, larger basins demand larger T values. By the way, in simulated annealing, the proper temperature parameter to escape in a rapid manner depends also on the depth of the attractor, a larger value is needed on the leftmost part of the landscape with respect to the rightmost part

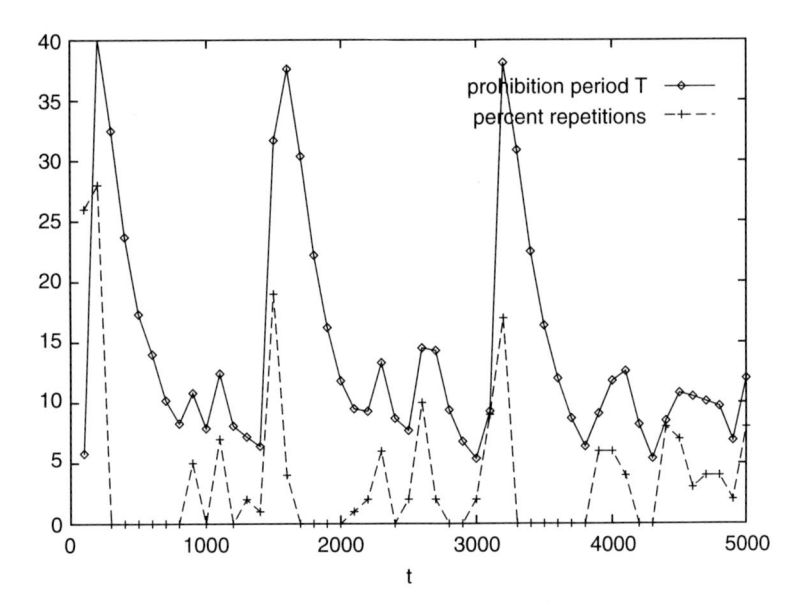

Fig. 4.11 Dynamics of the prohibition period T on a QAP task

Let us note that T is not fixed during the search, but is determined in a dynamic way depending on the *local structure* of the search space. This is particularly relevant for "inhomogeneous" tasks, where the statistical properties of the search space vary widely in the different regions (in these cases a fixed T would be inappropriate).

An example of the behavior of T during the search is illustrated in Fig. 4.11, for a Quadratic Assignment Problem task [33]. T increases in an exponential way when repetitions are encountered; it decreases in a gradual manner when repetitions disappear.

4.2.1 The Escape Mechanism

The basic tabu mechanism based on prohibitions is not sufficient to avoid long cycles (e.g., for binary strings of length L, T must be less than the length of the string, otherwise all moves are eventually prohibited, and therefore cycles longer than $2 \times L$ are still possible). In addition, even if "limit cycles" (endless cyclic repetitions of a given set of configurations) are avoided, the first reactive mechanism is not sufficient to guarantee that the search trajectory is not confined in a limited region of the search space. A "chaotic trapping" of the trajectory in a limited portion of the search space is still possible (the analogy is with *chaotic attractors* of dynamical systems, where the trajectory is confined in a limited portion of the space, although a limit cycle is not present).

For both reasons, to increase the robustness of the algorithm a second more radical diversification step (*escape*) is needed. The escape phase is triggered when too many configurations are repeated too often [33]. A simple escape action consists of a number of random steps executed after starting from the current configuration, possibly with a bias toward steps that bring the trajectory away from the current search region.

With a stochastic escape, one can easily obtain the *asymptotic convergence* of RTS. In a finite-cardinality search space, escape is activated infinitely often: if the probability for a point to be reached after escaping is different from zero for all points, eventually all points will be visited – clearly including the globally optimal points. The detailed investigation of the asymptotic properties and finite-time effects of different escape routines to enforce long-term diversification is an open research area.

4.2.2 Applications of Reactive Tabu Search

RTS has been used for widely different optimization tasks; let us mention some examples in the following. The experimental results in the seminal paper are on the 0–1 Knapsack Problem and on the Quadratic Assignment Problem [33, 34]. Applications to neural networks training are discussed in [36], leading also to special-purpose hardware realizations [8].

Vehicle routing applications are considered in [65, 202], unmanned aerial reconnaissance simulations in [221]. Graph partitioning is studied in [23], which compares different randomized and greedy strategies, with a description of the relationship between prohibitions and diversifications and between prohibition-based search and the original Kernighan-Lin algorithm.

Sensor selection in active structural acoustic control problems is studied in [160]. Methods based on hybrid combinations between tabu search and reactive GRASP are used in [84] for the single source capacitated plant location problem. Real-time dispatch of trams in storage yards is considered in [268].

Maximum satisfiability and constraint satisfaction problems are solved in [28–30]. The constraint satisfaction problem as a general problem solver is considered in [198], which discusses different rules for incrementing and decrementing the prohibition value.

The Maximum Clique Problem in graphs, a paradigmatic problem related to graph-based clustering, is considered in [31] with state-of-the-art results.

Electric power distribution and, in particular, service restoration in distribution power delivery systems is treated in [253].

Design optimization of N-shaped roof trusses, including also a population-base exploitation of the search history, is presented in [121]. Internet and telecommunications-related applications are described in [24, 98]. Variants of the multidimensional knapsack problem, related to the real-world issues of service level agreements and multimedia distributions, are cited in [128, 129].

An application in the field of continuous optimization is presented in [37] with the name Continuous RTS (C-RTS). The combinatorial optimization component, based on the RTS, locates the most promising "boxes," where starting points for the local minimizer are generated. Additional methods and comparisons are discussed in [62]. A different mixed RTS, combining discrete and continuous optimization, is used in [122] to design parts of vehicles (B-pillars) to maximize their resistance to roof crash.

A hybrid method combining RTS with ant colony optimization is proposed in [271] for constrained clique subgraph problems.

4.3 Implementation: Storing and Using the Search History

While the above classification deals with dynamical systems, a different classification is based on the detailed data structures used in the algorithms and on the consequent realization of the needed operations. Different data structures can possess widely different computational complexities so that attention should be spent on this subject before choosing a version of TS that is efficient on a particular problem.

Some examples of different implementations of the same TS dynamics are illustrated in Fig. 4.12. Strict-TS can be implemented through the reverse elimination method (REM) [81, 108], a term that refers to a technique for the storage and analysis of the ordered list of all moves performed throughout the search (called "running list"). The same dynamics can be obtained in all cases through standard *hashing* methods and storage of the configurations [33, 270], or, for the case of a search space consisting of binary strings, through the *radix tree* (or "digital tree") technique [33]. Hashing is an old tool in computer science: Different hashing functions – possibly with incremental calculation – are available for different domains [76]. REM is not applicable to all problems (the "sufficiency property" must be satisfied [108]), in addition its computational complexity per iteration is proportional to the number of iterations executed, whereas the average complexity obtained through incremental hashing is $O(1)$, a small and constant number of operations. The worst-case complexity per iteration obtained with the radix tree technique is proportional to the number of bits in the binary strings, and constant with respect to the iteration. If the memory usage is considered, both REM and approximated hashing use $O(1)$ mem-

4.3 Implementation: Storing and Using the Search History

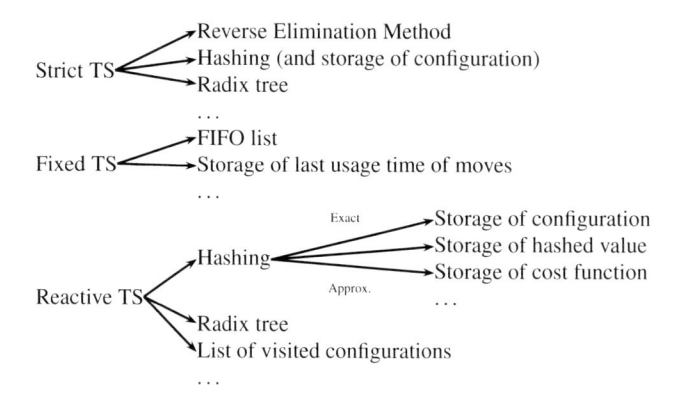

Fig. 4.12 The same search trajectory can be obtained with different data structures

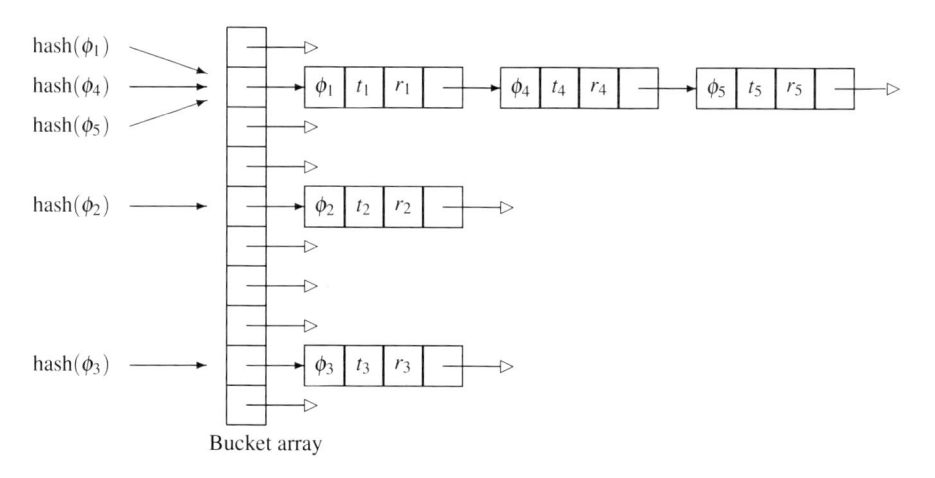

- Strict TS
  - Reverse Elimination Method
  - Hashing (and storage of configuration)
  - Radix tree
  - ...
- Fixed TS
  - FIFO list
  - Storage of last usage time of moves
  - ...
- Reactive TS
  - Hashing
    - Exact
      - Storage of configuration
      - Storage of hashed value
      - Storage of cost function
    - Approx.
    - ...
  - Radix tree
  - List of visited configurations
  - ...

Fig. 4.13 Open hashing scheme: items (configuration, or compressed hashed value, etc.) are stored in "buckets." The index of the bucket array is calculated from the configuration

ory per iteration, the actual number of bytes stored can be less for REM, because only changes (moves) and not configurations are stored.

Trivially, fixed-TS (alternative terms [107, 108] are simple-TS, static tabu search – STS – or tabu navigation method) can be realized with a first-in first-out list where the prohibited moves are located (the "tabu list"), or by storing in an array the last usage time of each move and by using (4.4).

Reactive-TS can be implemented through a simple list of visited configurations, or with more efficient hashing or radix tree techniques. At a finer level of detail, hashing can be realized in different ways. If the entire configuration is stored (see also Fig. 4.13), an exact answer is obtained from the memory lookup operation (a repetition is reported if and only if the configuration has been visited before). On the contrary, if a "compressed" item is stored, like a hashed value of a limited length derived from the configuration, the answer will have a limited probability of error (a repetition can be reported even if the configuration is new, because the compressed items are equal by chance – an event called "collision"). Experimentally, small collision probabilities do not have statistically significant effects, and hashing versions that need only a few bytes per iteration can be used.

4.3.1 Fast Algorithms for Using the Search History

The storage and access of the past events is executed through the well-known *hashing* or radix-tree techniques in a CPU time that is approximately constant with respect to the number of iterations. Therefore, the overhead caused by the use of the history is negligible for tasks requiring a nontrivial number of operations to evaluate the cost function in the neighborhood.

An example of a memory configuration for the hashing scheme is shown in Fig. 4.13. From the current configuration *phi*, one obtains an index into a "bucket array." The items (configuration or hashed value or derived quantity, last time of visit, total number of repetitions) are then stored in linked lists starting from the indexed array entry. Both storage and retrieval require an approximately constant amount of time if (i) the number of stored items is not much larger than the size of the bucket array and (ii) the hashing function scatters the items with a uniform probability over the different array indices. More precisely, given a hash table with m slots that stores n elements, a load factor $\alpha = n/m$ is defined. If collisions are resolved by chaining, searches take $O(1+\alpha)$ time, on average.

4.3.2 Persistent Dynamic Sets

Persistent dynamic sets are proposed to support memory-usage operations in history-sensitive heuristics in [20, 22]. Ordinary data structures are *ephemeral* [87], meaning that when a change is executed the previous version is destroyed. Now, in many contexts such as computational geometry, editing, implementation of very high-level programming languages, and, last but not least, the context of history-based heuristics, multiple versions of a data structure must be maintained and accessed. In particular, in heuristics one is interested in *partially persistent* structures, where all versions can be accessed but only the newest version (the *live* nodes) can be modified. A review of ad hoc techniques for obtaining persistent data structures is given in [87] that is dedicated to a systematic study of persistence, continuing the previous work of [204].

4.3.2.1 Hashing Combined with Persistent Red-Black Trees

The basic observation is that, because TS is based on local search, configuration $X^{(t+1)}$ differs from configuration $X^{(t)}$ only because of the addition or subtraction of a single index (a single bit is changed in the string). It is therefore reasonable to expect that more efficient techniques can be devised for storing *a trajectory of chained configurations* than for storing arbitrary states. The expectation is indeed true, although the techniques are not for beginners. You are warned, proceed only if not scared by advanced data structures.

Let us define the operations INSERT(i) and DELETE(i) for inserting and deleting a given index i from the set. As cited above, configuration X can be considered as a set of indices in $[1, L]$ with a possible realization as a balanced red-black tree; see [39, 118] for two seminal papers about red-black trees. The binary string can be immediately obtained from the tree by visiting it in symmetric order, in time $O(L)$. INSERT(i) and DELETE(i) require $O(\log L)$ time, while at most a single node of the tree is allocated or deallocated at each iteration. Rebalancing the tree after insertion or deletion can be done in $O(1)$ rotations and $O(\log L)$ color changes [250]. In addition, the amortized number of color changes per update is $O(1)$; see for example [176].

Now, the REM method [107, 108] is closely reminiscent of a method studied in [204] to obtain partial persistence, in which the entire update sequence is stored and the desired version is rebuilt from scratch each time an access is performed, while a systematic study of techniques with better space–time complexities is present in [87, 222]. Let us now summarize from [222] how a partially persistent red-black tree can be realized. An example of the realizations that we consider is presented in Fig. 4.14.

The trivial way is that of keeping in memory all copies of the ephemeral tree (see the top part of Fig. 4.14), each copy requiring $O(L)$ space. A smarter realization is based on *path copying*, independently proposed by many researchers; see [222] for references. Only the path from the root to the nodes where changes are made is copied: A set of search trees is created, one per update, having different roots but sharing common subtrees. The time and space complexities for INSERT(i) and DELETE(i) are now of $O(\log L)$.

The method that we will use is a space-efficient scheme requiring only linear space proposed in [222]. The approach avoids copying the entire access path each time an update occurs. To this end, each node contains an additional "extra" pointer (beyond the usual left and right ones) with a time stamp. When attempting to add a pointer to a node, if the extra pointer is available, it is used and the time of the usage is registered. If the extra pointer is already used, the node is copied, setting the initial left and right pointers of the copy to their latest values. In addition, a pointer to the copy is stored in the last parent of the copied node. If the parent has already used the extra pointer, the parent too, is copied. Thus copying proliferates through successive ancestors until the root is copied or a node with a free extra pointer is encountered. Searching the data structure at a given time t in the past is easy: After starting from the appropriate root, if the extra pointer is used the pointer to follow from a node is determined by examining the time stamp of the extra pointer and following it if and only if the time stamp is not larger than t. Otherwise, if the extra pointer is not used, the normal left–right pointers are considered. Note that the pointer direction (left or right) does not have to be stored: Given the search

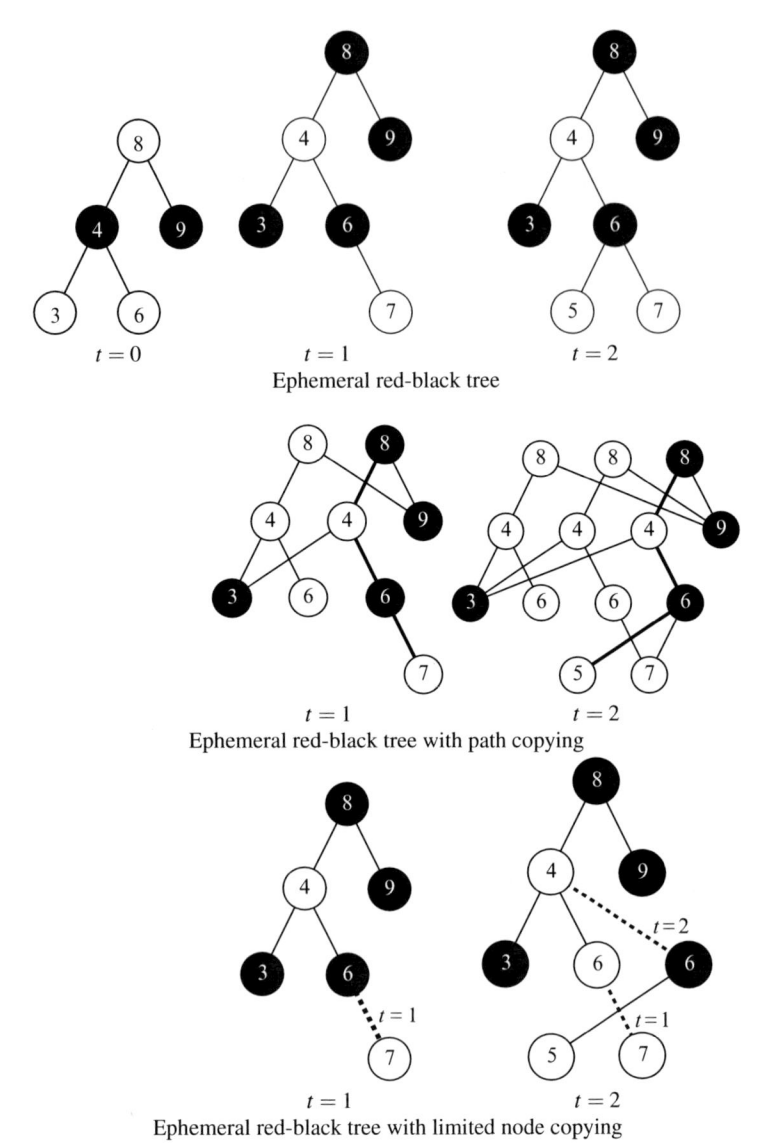

Fig. 4.14 How to obtain a partially persistent red-black tree from an ephemeral one (*top*), containing indices 3,4,6,8,9 at t = 0, with subsequent insertion of 7 and 5. Path copying (*middle*), with thick lines marking the copied part. Limited node copying (*bottom*) with dashed lines denoting the "extra" pointers with time stamp

tree property, it can be derived by comparing the indices of the children with that of the node. In addition, colors are needed only for the most recent (live) version of the tree. In Fig. 4.14, NULL pointers are not shown, colors are correct only for the live tree (the nodes reachable from the rightmost root), and extra pointers are dashed and time-stamped.

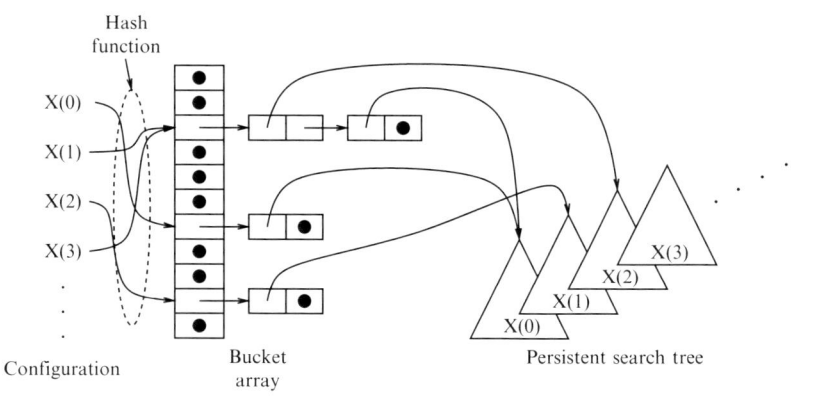

4.3 Implementation: Storing and Using the Search History

Fig. 4.15 Open hashing scheme with persistent sets: A pointer to the appropriate root for configuration $X^{(t)}$ in the persistent search tree is stored in a linked list at a "bucket." Items on the list contain satellite data. The index of the bucket array is calculated from the configuration through a hashing function

The worst-case time complexity of INSERT(i) and DELETE(i) remains $O(\log L)$, but the important result derived in [222] is that the amortized space cost per update operation is $O(1)$. Let us recall that the total amortized space cost of a sequence of updates is an upper bound on the actual number of nodes created.

Let us now consider the context of history-based heuristics. Contrary to the popular usage of persistent dynamic sets to search past versions at a specified time t, one is interested in checking whether a configuration has already been encountered in the previous history of the search, at *any* iteration.

A convenient way of realizing a data structure supporting X-SEARCH(X) is to combine *hashing* and *partially persistent dynamic sets*; see Fig. 4.15. From a given configuration X, an index into a "bucket array" is obtained through a hashing function, with a possible incremental evaluation in time $O(1)$. Collisions are resolved through chaining: Starting from each bucket header there is a linked list containing a pointer to the appropriate root of the persistent red-black tree and satellite data needed by the search (time of configuration, number of repetitions).

As soon as configuration $X^{(t)}$ is generated by the search dynamics, the corresponding persistent red-black tree is updated through INSERT(i) or DELETE(i). Let us now describe X-SEARCH($X^{(t)}$): The hashing value is computed from $X^{(t)}$ and the appropriate bucket searched. For each item in the linked list the pointer to the root of the past version of the tree is followed and the old set is compared with $X^{(t)}$. If the sets are equal, a pointer to the item on the linked list is returned. Otherwise, after the entire list has been scanned with no success, a NULL pointer is returned.

In the last case a new item is linked in the appropriate bucket with a pointer to the root of the live version of the tree (X-INSERT(X,t)). Otherwise, the last visit time t is updated and the repetition counter is incremented.

After collecting the above-cited complexity results, and assuming that the bucket array size is equal to the maximum number of iterations executed in the entire search, it is straightforward to conclude that each iteration of *reactive-TS* requires $O(L)$ average-case time and $O(1)$ amortized space for storing and retrieving the past configurations and for establishing prohibitions.

In fact, both the hash table and the persistent red-black tree require $O(1)$ space (amortized for the tree). The worst-case time complexity per iteration required to update the current $X^{(t)}$ is $O(\log L)$, the average-case time for searching and updating the hashing table is $O(1)$ (in detail, searches take time $O(1+\alpha)$, α being the load factor, in our case upper bounded by 1). The time is therefore dominated by that required to compare the configuration $X^{(t)}$ with that obtained through X-SEARCH$(X^{(t)})$, i.e., $O(L)$ in the worst case. Because $\Omega(L)$ time is needed during the neighborhood evaluation to compute the f values, the above complexity is optimal for the considered application to history-based heuristics.

Chapter 5
Reacting on the Objective Function

> *How to Do a Proper Push-Up. Push-ups aren't just for buff army trainees; they are great upper body, low-cost exercise. Here's the proper way to do them anywhere.*
> *(Advertisement in the web)*

5.1 Dynamic Landscape Modifications to Influence Trajectories

This chapter considers reactive modification of the objective function in order to support appropriate diversification of the search process. Contrary to the prohibition-based techniques of Chap. 4 the focus is not that of pushing the search trajectory away from a local optimum though explicit and direct prohibitions but on modifying the objective function so that previous promising areas in the solution space appear less favorable, and the search trajectory will be gently pushed to visit new portions of the search space. To help the intuition, see also Fig. 5.1, one may think about pushing up the search landscape at a discovered local minimum, so that the search trajectory will flow into neighboring attraction basins. For a physical analogy, think about you sitting in a tent while it is raining outside. A way to eliminate dangerous pockets of water stuck on flat convex portions is to gently push the tent fabric from below until gravity will lead water down.

As with many algorithmic principles, it is difficult to pinpoint a seminal paper in this area. The literature about stochastic local search for the satisfiability (SAT) problem is of particular interest. Different variations of local search with randomness techniques have been proposed for satisfiability and maximum satisfiability (MAX-SAT) starting from the late 1980s, for some examples see [117, 231], and the updated review of [146]. These techniques were in part motivated by previous applications of "min-conflicts" heuristics in the area of artificial intelligence; see for example [116] and [185].

Before arriving at the objective function modifications, let us summarize the influential algorithm GSAT [231]. It consists of multiple runs of LS^+ local search, each one consisting of a number of iterations that is typically proportional to the problem dimension n. Let f be the number of satisfied clauses. At each iteration of LS^+, a bit that maximizes Δf is chosen and flipped, even if Δf is negative, i.e., after flipping the bit the number of satisfied clauses decreases.

The algorithm is briefly summarized in Fig. 5.2. A certain number of tries (*MAX-TRIES*) is executed, where each try consists of a number of iterations

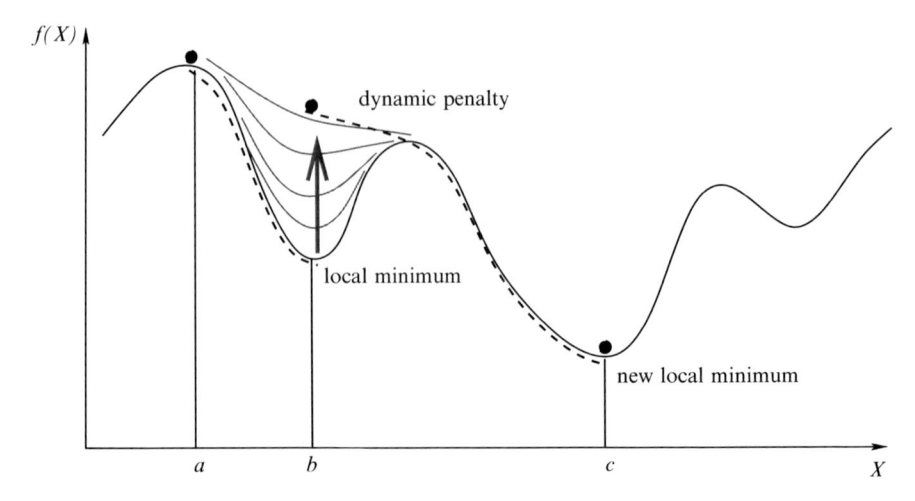

Fig. 5.1 Transformation of the objective function to gently push the solution out of a given local minimum

```
GSAT-WITH-WALK
1       for i ← 1 to MAX-TRIES
2           X ← random truth assignment
3           for j ← 1 to MAX-FLIPS
4               if RAND(0,1)< p then
5                   var ← any variable occurring in some unsatisfied clause
6               else
7                   var ← any variable with largest Δf
8               FLIP(var)
```

Fig. 5.2 The "GSAT-with-walk" algorithm. RAND(0, 1) generates random numbers in the range [0, 1]

(*MAX-FLIPS*). At each iteration, a variable is chosen by two possible criteria and then flipped by the function FLIP, i.e., x_i becomes equal to $(1 - x_i)$. One criterion, active with *noise* probability p, selects a variable occurring in some unsatisfied clause with uniform probability over such variables; the other one is the standard method based on the function f given by the number of satisfied clauses. For a generic move μ applied at iteration t, the quantity $\Delta_\mu f$ (or Δf for short) is defined as $f(\mu X^{(t)}) - f(X^{(t)})$. The straightforward book-keeping part of the algorithm is not shown. In particular, the best assignment found during all trials is saved and reported at the end of the run. In addition, the run is terminated immediately if an assignment is found that satisfies all clauses. Different noise strategies to escape from attraction basins are added to GSAT in [229, 230], In particular, the GSAT-with-walk algorithm.

The breakout method suggested in [190] for the constraint satisfaction problem measures the cost as the sum of the weights associated to the violated constraints (to the nogoods). Each weight is one at the beginning, at a local minimum the weight of each nogood is increased by one until one escapes from the given local minimum (a breakout occurs).

5.1 Dynamic Landscape Modifications to Influence Trajectories

Clause weighting has been proposed in [228] in order to increase the effectiveness of GSAT for problems characterized by strong asymmetries. In this algorithm a positive weight is associated to each clause to determine how often the clause should be counted when determining which variable to flip. The weights are dynamically modified during problem solving and the qualitative effect is that of "filling in" local optima while the search proceeds. Clause-weighting and the breakout technique can be considered as "reactive" techniques where a repulsion from a given local optimum is generated in order to induce an escape from a given attraction basin. The local adaptation is clear: Weights are increased until the original local minimum disappears, and therefore the current weights depend on the local characteristic of a specific local minimum point.

In detail, a weight w_i is associated to each clause, and the guiding evaluation function becomes not a simple count of the satisfied clauses but a sum of the corresponding weights. New parameters are introduced and therefore new possibilities for tuning the parameters based on feedback from preliminary search results. The algorithm in [227] suggests a different way to use weights to encourage more priority on satisfying the "most difficult" clauses. One aims at *learning how difficult a clause is to satisfy*. These hard clauses are identified as the ones that remain unsatisfied after a try of local search descent followed by plateau search. Their weight is increased so that future runs will give them more priority when picking a move. More algorithms based on the same weighting principle are proposed in [99, 100], where clause weights are updated after each flip: The reaction from the unsatisfied clauses is now immediate as one does not wait until the end of a try (weighted GSAT or WGSAT). If weights are only increased, after some time their size becomes large and their relative magnitude will reflect the overall statistics of the SAT instance, more than the local characteristics of the portion of the search space where the current configuration lies. To combat this problem, two techniques are proposed in [100], either *reducing* the clause weight when a clause is satisfied, or storing the weight increments that took place recently, which is obtained by a weight decay scheme (each weight is reduced by a factor ϕ before updating it). Depending on the size of the increments and decrements, one achieves "continuously weakening incentives not to flip a variable" instead of the strict prohibitions of TS (see Chap. 4). The second scheme takes the *recency of moves* into account, the implementation is through a weight decay scheme updating so that each weight is reduced before a possible increment by δ if the clause is not satisfied:

$$w_i \leftarrow \phi\, w_i + \delta,$$

where one introduces a decay rate ϕ and a "learning rate" δ. A faster decay (lower ϕ value) will limit the temporal extension of the context and imply a faster forgetting of old information. The effectiveness of the weight decay scheme is interpreted by the authors as "learning the best way to conduct local search by discovering the hardest clauses relative to recent assignments." A critique of some *warping* effects that a clause-weighting dynamic local search can create on the fitness surface is present in [251]: In particular let us note that the fitness surface is changed in a global way after encountering a local minimum. Points that are very far from the local minimum, but that share some of the unsatisfied clauses, will also see their values changed. This does not correspond to the naive "push-up" picture where only the area close to a specific local minimum is raised, and the effects on the overall search dynamics are far from simple to understand.

A more recent proposal of a dynamic local search (DLS) for SAT is in [252]. The authors start from the exponentiated sub-gradient (ESG) algorithm [225], which alternates search phases and weight updates, and develop a scheme with low-time complexity of its search steps: Scaling and Probabilistic Smoothing (SAPS). Weights of satisfied clauses are multiplied by α_{sat}, while weights of unsatisfied clauses are multiplied by α_{unsat}, then all weights are smoothed toward their mean $\bar{w}$: $w \leftarrow w\rho + (1-\rho)\bar{w}$. A *reactive version* of SAPS (RSAPS) is then introduced that adaptively tunes one of the algorithm's important parameters.

5.1.1 Adapting Noise Levels

Up to now we have seen how to modify the objective function in a dynamic manner depending on the search history and current state in order to deviate the search trajectory away from local optimizer. Another possibility to reach similar (stochastic) deviations of the trajectory is by adding a controlled amount of randomized movements. *Clinamen* is the name the philosopher Lucretius gave to the spontaneous microscopic swerving of atoms from a vertical path as they fall, considered in discussions of possible explanation for free will. A kind of algorithmic clinamen can be used to influence the diversification of an SLS technique. This part is considered in this chapter because the net effect on the trajectory can be similar to that obtained by modifying the objective function. A related usage of "noise" is also considered in Chap. 3 about simulated annealing, in which upward moves are accepted with a probability depending on a temperature parameter.

An opportunity for self-adaptation considering different amounts of randomness is given by *adaptive noise* mechanism for WalkSAT. In WalkSAT [229], one repeatedly flips a variable in an unsatisfied clause. If there is at least one variable that can be flipped without breaking already satisfied clauses, one of them is flipped. Otherwise, a noise parameter p determines if a random variable is flipped, or if a greedy step is executed, with probability $(1-p)$, favoring minimal damage to the already satisfied clauses.

In [181] it appears that appropriate noise settings achieve a good balance between the greedy "steepest descent" component and the exploration of other search areas away from already considered attractors. Parameters with a diversifying effect similar to the noise in WalkSAT are present in many techniques, for example, in prohibition-based techniques, see Chap. 4. The work in [181] considers this generalized notion of a noise parameter and suggests tuning the proper noise value for a specific instance by testing different settings through a preliminary series of short runs. Furthermore, the suggested statistics to monitor, which is closely related to the algorithm performance, is the *ratio* between the average final values obtained at the end of the short runs and the variance of the f values over the runs. Quite consistently, the best noise setting corresponds to the one leading to the lowest empirical ratio increased by about 10%. Faster tuning can be obtained if the examination of a predefined series of noise values is substituted with a faster adaptive search that considers a smaller number of possible values; see [208], which uses Brent's method [55]. An adaptive noise scheme is also proposed in [136], where the noise setting p is dynamically adjusted based on search progress. Higher noise levels are determined in a reactive manner if and only if there is evidence of search stagnation. In detail, if too many steps elapse since the last improvement, the noise value is increased, while it is gradually decreased if evidence of stagnation disappears. A different approach based on optimizing the noise setting on a given instance prior to the actual search process (with a fixed noise setting) is considered in [208].

5.1.2 Guided Local Search

While we concentrated on the SAT problem above, a similar approach has been proposed with the term of Guided Local Search (GLS) [260, 261] for other applications. GLS aims at enabling intelligent search schemes that exploit problem- and search-related information to guide a local search algorithm in a search space. Penalties depending on solution features are introduced and dynamically manipulated to distribute the search effort over the regions of a search space.

Let us stop for a moment with an historical digression to show how many superficially distinct concepts are in fact deeply related. Inspiration for GLS comes from a previously proposed neural net algorithm (GENET) [263] and from tabu search [107], simulated annealing [161], and tunneling [170]. The use of "neural networks" for optimization consists of setting up *a dynamical system whose attractors correspond to good solutions of the optimization problem*. Once the dynamical system paradigm is in the front stage, it is natural to use it not only to search for but also to escape from local minima. According to the authors [259], GENET's mechanism for escaping resembles *reinforcement learning* [17]: Patterns in a local minimum are stored in the constraint weights and are discouraged to appear thereafter. GENET's learning scheme can be viewed as a method to *transform the objective function so that a local minimum gains an artificially higher value*. Consequently, local search will be able to leave the local minimum state and search other parts of the space. In tunneling algorithms [170], the modified objective function is called the tunneling function. This function allows local search to explore states that have higher costs around or further away from the local minimum, while aiming at nearby states with lower costs. In the framework of continuous optimization, similar ideas have been rediscovered multiple times. Rejection-based stochastic procedures are presented in [11, 170, 200]. Citing from a seminal paper [170], one combines "a minimization phase having the purpose of lowering the current function value until a local minimizer is found and a tunneling phase that has the purpose of finding a point ... such that when employed as starting point for the next minimization phase, the new stationary point will have a function value no greater than the previous minimum found." The "strict" prohibitions of tabu search become "softer" penalties in GLS, which are determined by *reaction to feedback from the local optimization heuristic under guidance* [261].

A complete GLS scheme [261] defines appropriate solution features f_i, for example, the presence of an edge in a TSP path, and combines three ingredients:

Feature penalties p_i to diversify the search away from already-visited local minima (the *reactive* part)

Feature costs c_i to account for the a priori promise of solution features (for example, the edge cost in TSP)

A neighborhood activation scheme   depending on the current state.

The *augmented cost function* $h(X)$ is defined as:

$$h(X) = f(X) + \lambda \sum_i p_i\, I_i(X), \quad (5.1)$$

where $I_i(X)$ is an indicator function returning 1 if feature i is present in solution X, 0 otherwise. The augmented cost function is used by local search instead of the original function.

Penalties are zero at the beginning: There is no need to escape from local minima until they are encountered! Local minima are then the "learning opportunities" of GLS: When a local minimum of h is encountered, the augmented cost function is modified by updating the penalties p_i. One considers all features f_i present in the local minimum solution X' and increments by one the penalties that maximize:

$$I_i(X')\, \frac{c_i}{1+p_i}. \quad (5.2)$$

The above mechanism kills more birds with one stone. First a higher cost c_i, and therefore an inferior a priori desirability for feature f_i in the solution, implies a higher tendency to be penalized. Second, the penalty p_i, which is also a counter of how many times a feature has been penalized, appears at the denominator, and therefore discourages penalizing features that have been penalized many times in the past. If costs are comparable, the net effect is that penalties tend to alternate between different features present in local minima.

GLS is usually combined with "fast local search" FLS. FLS includes implementation details that speedup each step but do not impact the dynamics and do not change the search trajectory, for example, an incremental evaluation of the h function, but it also includes qualitative changes in the form of *subneighborhoods*. The entire neighborhood is broken down into a number of small subneighborhoods. Only active subneighborhoods are searched. Initially all of them are active, then, if no improving move is found in a subneighborhood, it becomes inactive. Depending on the move performed, a number of subneighborhoods are activated, where one expects improving moves to occur as a result of the move just performed. For example, after a feature is penalized, the subneighborhood containing a move eliminating the

5.1 Dynamic Landscape Modifications to Influence Trajectories

feature from the solution is activated. The mechanism is equivalent to prohibiting examination of the inactive moves, in a tabu search spirit. As an example, in TSP one has a *subneighborhood* per city, containing all moves exchanging edges where at least one of the edges terminates at the given city. After a move is performed, all *subneighborhoods* corresponding to cities at the ends of the edges involved in the move are activated, to favor a chain of moves involving more cities.

While the details of *subneighborhood* definition and update are problem-dependent, the lesson learned is that much faster implementations can be obtained by avoiding a brute-force evaluation of the neighborhood, the motto is "evaluate only a subset of neighbors where you expect improving moves." In addition to a faster evaluation per search step, one obtains a possible additional diversification effect related to the implicit prohibition mechanism. This technique to speedup the evaluation of the neighborhoods is similar to the "don't look bits" method in [41]. One flag bit is associated to every node, and if its value is 1 the node is not considered as a starting point to find an improving move. Initially all bits are zero; then if an improving move could not be found starting at node i, the corresponding bit is set. The bit is cleared as soon as an improving move is found that inserts an edge incident to node i.

The parameter λ in (5.1) controls the importance of penalties with respect to the original cost function: A large λ implies a large diversification away from previously visited local minima. A reactive determination of the parameter λ is suggested in [261].

While the motivations of GLS are clear, the interaction between the different ingredients causes a somewhat complicated dynamics. Let us note that different units of measure for the cost in (5.1) can impact the dynamics, something that is not particularly desirable: if the cost of edge in TSP is measured in kilometers, the dynamics is not the same as if the cost is measured in millimeters. Furthermore, the definition of costs c_i for a general problem is not obvious and the consideration of the "costs" c_i in the penalties in a way duplicates the explicit consideration of the real problem costs in the original function f. In general, when penalties are added and modified, a desired effect (minimal required diversification) is obtained *indirectly* by modifying the objective function and therefore by possibly causing *unexpected effects*, like new spurious local minima, or shadowing of promising yet-unvisited solutions. For example, an unexplored local minimum of f may not remain a local minimum of h and therefore it may be skipped by modifying the trajectory.

A penalty formulation for TSP, including memory-based trap-avoidance strategies is proposed in [262]. One of the strategies avoids visiting points that are close to points visited before, a generalization of the STRICT-TS strategy; see Chap. 4. A recent algorithm with an *adaptive clause weight redistribution* is presented in [145]; it adopts resolution-based preprocessing and reactive adaptation of the total amount of weight to the degree of stagnation of the search. Stagnation is identified after a long sequence of flips without improvement, long periods of stagnation will produce "oscillating phases of weight increase and reduction."

5.2 Eliminating Plateaus by Looking Inside the Problem Structure

In the above presentation we considered modifications of the objective functions in order to modify the trajectory dynamics to escape from already-visited attractors. We now consider modifications that have a different purpose: That of eliminating *plateaus*. A *plateau* is a situation where one has a local minimum, but some neighbors have *the same f* value. By moving on a *plateau* one keeps a good starting point at a low f value, with the usual hope to eventually reach an improving move. But large plateaus are always "embarrassing:" one is stuck at a flat desert looking for water, no sun to give a direction. If one is not careful a lot of time can be spent looking around, maybe retracing the previous steps, with the thirst growing harder and harder. One would like some hints about a promising direction to take, maybe some humid mold so that water will get closer and closer by following the humidity gradient. Coming back from mirages to algorithms, one aims at breaking the ties among seemingly equivalent solutions, equivalent when considering only the f values. In particular, it may be the case that, while f is constant, some internal changes in the solution structure will eventually favor the discovery of an improving move. For example, in the MAX-SAT problem, even if the number of satisfied clauses remains the same, the amount of *redundancy* in their satisfaction (the number of different literals that make a clause satisfied) may pave the way to eventually flipping a variable that is redundant to satisfy some already-satisfied clauses in order to satisfy a new one. Aiming at a redundant satisfaction eliminates the embarrassment in selecting among seemingly similar situations and favors an improvement after a smaller number of steps than those required by a random-walk on the plateau.

5.2.1 Nonoblivious Local Search for SAT

In the design of efficient approximation algorithms for MAX–SAT, an approach of interest is based on the use of *nonoblivious functions* independently introduced in [5] and in [235]. Let us consider the classical local search algorithm LS for MAX–SAT, here redefined as *oblivious* local search (LS-OB). Now, a different type of local search can be obtained by using a *different* objective function to direct the search, i.e., to select the best neighbor at each iteration. Local optima of the standard objective function f are not necessarily local optima of the different objective function. In this event, the second function causes an *escape* from a given local optimum. Interestingly enough, suitable *nonoblivious* functions f_{NOB} improve the performance of LS if one considers both the worst-case performance ratio and, as it has been shown in [27], the actual average results obtained on benchmark instances.

Let us introduce the notation and mention a theoretical result for MAX–2–SAT. Given an assignment X, let S_i denote the set of clauses in the given task in which exactly i literals are true and let $w(S_i)$ denote the cardinality of S_i. In

5.2 Eliminating Plateaus by Looking Inside the Problem Structure

addition, a d-neighborhood of a given truth assignment is defined as the set of all assignment where the value of at most d variables is changed. The theoretically-derived nonoblivious function for MAX–2–SAT is:

$$f_{NOB}(X) = \frac{3}{2}w(S_1) + 2w(S_2).$$

Theorems 7–8 of [235] state that the performance ratio for any LS-OB algorithm with a d-neighborhood for MAX–2–SAT is $2/3$ for any $d = o(n)$, while nonoblivious local search with an 1-neighborhood achieves a performance ratio $3/4$. Therefore, LS-NOB improves considerably the performance ratio even if the search is restricted to a much smaller neighborhood. In general, LS-NOB achieves a performance ratio $1 - \frac{1}{2^k}$ for MAX–k–SAT. The oblivious function for MAX–k–SAT is of the form:

$$f_{NOB}(X) = \sum_{i=1}^{k} c_i w(S_i)$$

and the above given performance ratio is obtained if the quantities $\Delta_i = c_{i+1} - c_i$ satisfy:

$$\Delta_i = \frac{1}{(k-i+1)\binom{k}{i-1}} \left[\sum_{j=0}^{k-i} \binom{k}{j} \right].$$

Because the positive factors c_i that multiply $w(S_i)$ in the function f_{NOB} are strictly increasing with i, the approximations obtained through f_{NOB} tend to be characterized by a *"redundant" satisfaction of many clauses*. Better approximations, at the price of a limited number of additional iterations, can be obtained by a two-phase local search algorithm (NOB&OB): After a random start, f_{NOB} guides the search until a local optimum is encountered. As soon as this happens, a second phase of LS is started where the move evaluation is based on f. A further reduction in the number of unsatisfied clauses can be obtained by a "plateau search" phase following NOB&OB: The search is continued for a certain number of iterations after the local optimum of OB is encountered, by using LS$^+$, with f as guiding function.

Let us note that a similar proposal to define an objective function that considers "how strongly clauses are satisfied" has been proposed later in [224], coupled with a multiplicative reweighting of unsatisfied clauses ("smoothed descent and flood"). According to the authors, "additive updates do not work very well because clauses develop large weight differences over time, and this causes the update mechanism to lose its ability to rapidly adapt the weight profile to new regions of the search space." Again, the possibility to react rapidly to local characteristics is deemed of particular importance.

Chapter 6
Model-Based Search

> *The sciences do not try to explain, they hardly even try to interpret, they mainly make models. By a model is meant a mathematical construct which, with the addition of certain verbal interpretations, describes observed phenomena. The justification of such a mathematical construct is solely and precisely that it is expected to work.*
> *(Johann Von Neumann)*

6.1 Models of a Problem

In the previous chapters we concentrated on solving optimization problems by applying different flavors of the local search paradigm. Nonetheless, the same problems can be thought of from a more global point of view, leading to a different family of useful optimization techniques.

The main idea of model-based optimization is to create and maintain a *model* of the problem, which aims to provide some clues about the problem's solutions. If the problem is a function to be minimized, for instance, it is helpful to think of such model as a *simplified version* of the function itself. In more general settings, the model can be a probability distribution defining the estimated likelihood of finding a good quality solution at a certain point.

To solve a problem, we resort to the model in order to generate a candidate solution, and then check it. The result of the check shall be used to refine the model, so that the future generation is biased toward better and better candidate solutions. Clearly, for a model to be useful, it must provide as much information about the problem as possible, while being somehow "more tractable," in a computational or analytical sense, than the problem itself. The initial model can be created through prior knowledge or by uniformity assumptions.

Although memory-based techniques can be used in both discrete and continuous domains, the latter case better supports our intuition. In Fig. 6.1, a function (continuous line) must be minimized. An initial model (the dashed line) provides a prior probability distribution for the minimum. In case of no prior knowledge, a uniform distribution can be assumed. Based on this estimate, some candidate minima are generated (points a through d), and the corresponding function values are computed. The model is updated (dotted line) to take into account the latest findings: The global minimum is more likely to occur around c and d, rather than around a and b. Further model-guided generations and tests will improve the distribution: Eventually the region around the global minimum e shall be discovered and a high probability density will be assigned to its surroundings. The same example also

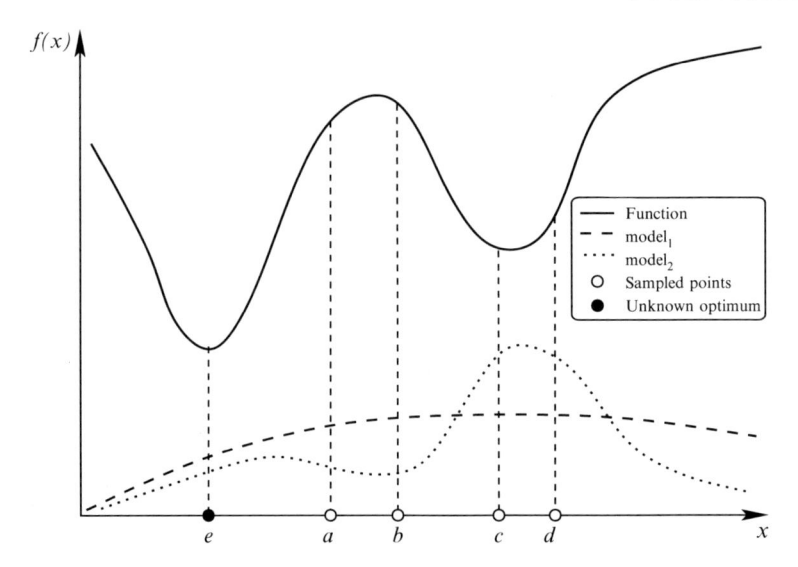

Fig. 6.1 Model-based search: One generates sample points from model$_1$ and updates the generative model to increase the probability for point with low cost values (see model$_2$). If one is not careful, the optimal point e runs the risk of becoming more and more difficult to generate

highlights a possible drawback of naïf applications of the technique: Assigning a high probability to the neighborhood of c and d could lead to a negligible probability of selecting a point near e, so the global minimum would never be discovered. It looks like the emphasis is on *intensification* of the search. This is why, in practice, the models are corrected to ensure a significant probability of generating points also in unexplored regions.

An alternative possibility would be to avoid the model altogether and to keep in memory all or a fraction of the previously tested points. This hypothesis reminds of search algorithms based on a dynamic population of sample points, also known as genetic or evolutionary algorithms; see Chap. 11 for more details. The informed reader will also notice a similarity to the machine learning subdivision between *instance-based* and *model-based* learning techniques; see for example [213].

Now that we are supported by intuition, let us proceed with the theoretical aspects. The discussion is inspired by the recent survey in [276]. The scheme of a model-based search approach is presented in Fig. 6.2. Represented entities are:

- A model used to generate sample solutions,
- The last samples generated,
- A memory containing previously accumulated knowledge about the problem (previous solutions and evaluations).

The process develops in an iterative way through a feedback loop where new candidates are generated by the model, and their evaluation – together with memory about past states – is used to improve the model itself in view of a new generation.

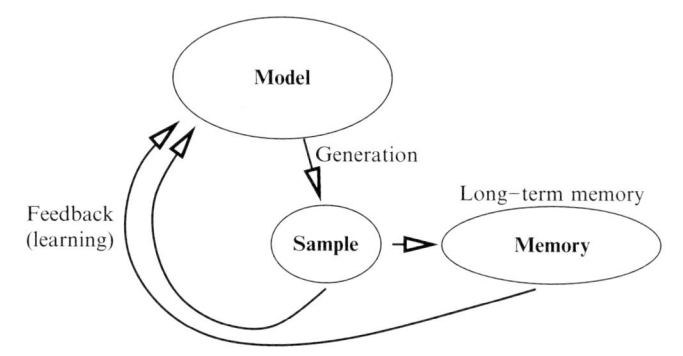

Fig. 6.2 Model-based architecture: A generative model is updated after learning from the last generated samples and the previous long-term memory

The design choices consist of defining a suitable generative model, and an appropriate learning rule to favor the generation of superior models in the future steps. A lot of computational complexity can lurk within the second issue, which is in itself an optimization task. In particular, one should avoid learning rules converging to local optima, as well as overtraining, which would hamper generalization.

6.2 An Example

Let us start from a simple model. The search space $\mathscr{X} = \{0,1\}^L$ is the set of all binary strings of length L, the generation model is defined by an L-tuple of parameters

$$p = (p_1, \ldots, p_L) \in [0,1]^L,$$

where p_i is the probability of producing 1 as the i-th bit of the string and every bit is independently generated. The motivation is to "remove genetics from the standard genetic algorithm" [14]: Instead of maintaining implicitly a statistic in a GA population, *statistics are maintained explicitly* in the vector (p_i).

The initial state of the model corresponds to indifference with respect to the bit values: $p_i = 0.5$, $i = 1, \ldots, L$. In the population-based incremental learning (PBIL) algorithm [14] the following steps are iterated:

1. Initialize p;
2. **repeat**:
3.   Generate a sample set S using the vector p;
4.   Extract a fixed number $\bar{S}$ of the best solutions from S;
5.   **for each** sample $s = (s_1, \ldots, s_n) \in \bar{S}$:
6. $p \leftarrow (1-\rho)p + \rho s,$

where ρ is a learning rate parameter regulating exploration vs. exploitation. A smaller ρ means that the effect of the last samples is reduced, so that a longer-term

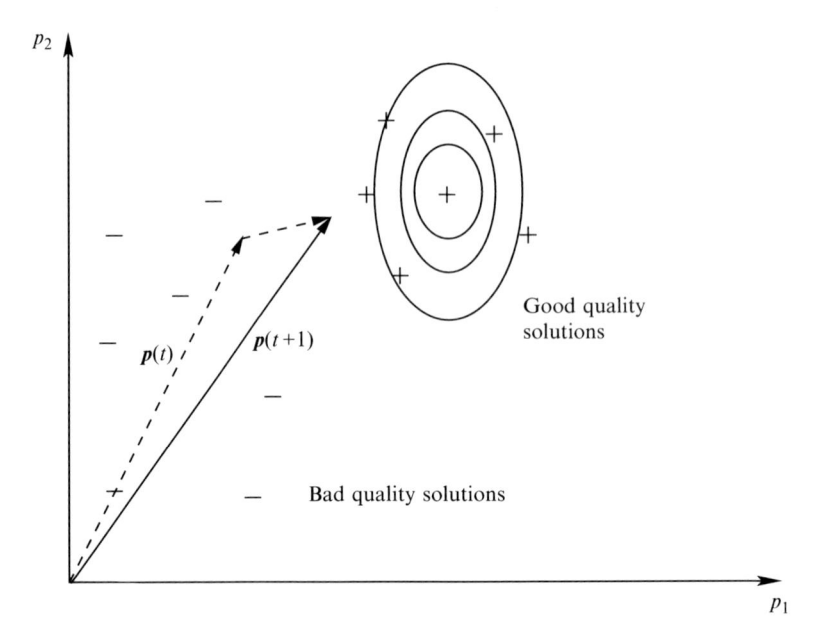

Fig. 6.3 PBIL: the "prototype" vector **p** gradually shifts toward good quality solutions (qualitative example in two dimensions)

memory about the past samples is kept in the model vector p. The moving vector p can be seen as representing a moving average of the best samples, a *prototype* vector placed in the middle of the cluster providing the best quality solutions. Parallel schemes exist in the neural networks and machine learning literature, for example, the update rule is similar to that used in learning vector quantization; see [127]. Variations include moving away from bad samples in addition to moving toward good ones. A schematic representation is shown in Fig. 6.3.

Although this simple technique shows results superior to GA on some benchmark tasks [14], the method has intrinsic weaknesses if the optimization landscape has a rich structure, for example, more than a single cluster, or complex configurations of the optimal positions corresponding to *dependencies among the individual bits*.

Estimation-of-distributions algorithms [192] (EDAs) have been proposed in the framework of evolutionary computation for modeling promising solutions in a probabilistic manner. As the name suggests, in EDAs a new population (a set of candidate solutions in the genetic algorithm terminology) is generated by sampling a probability distribution, which is estimated from a database containing selected individuals of previous generations. A survey in [209] considers population-based search algorithms based on probabilistic models to guide the exploration of the search space. A recent theoretical study in [272] evaluates the advantages of using higher-order statistics in EDAs. Two EDAs with two-tournament selection for discrete optimization problems are considered and the findings suggest that using higher order statistics can indeed improve the chance of finding the global optimum.

6.3 Dependent Probabilities

Estimates of probability densities for optimization in the form of pairwise conditional probabilities are studied in [49], a paper that also suggests a clear theoretical framework. Their MIMIC technique (mutual-information-maximizing input clustering) aims at estimating a probability density for points with value below a given threshold (remember that the function is to be *minimized*). The method aims at modeling the distribution p^θ, which is uniform over inputs X with $f(X) \leq \theta$, and is zero elsewhere: If this distribution is known and the best value $\hat{\theta} = \min_X f(X)$ is also known, a single sample from $p^{\hat{\theta}}$ would be sufficient to identify the optimizer.

Before proceeding with the method, a proper threshold θ is to be selected. In the absence of a priori knowledge, we can aim at tuning θ so that it permits a nontrivial estimation of p^θ. If θ is too low, no sample point will make it; if it is too large, all points are in. The choice in [49] is to *adapt* θ so that it is equal to a fixed Nth percentile of the data (so that a specific fraction of the sample points is below the threshold). If the model is appropriate, one aims at a *threshold-descent* behavior: The threshold will be high at the beginning, then the areas leading to lower and lower costs will be identified leading to a progressively lower threshold (a similar technique will be encountered for racing in Chap. 10). The algorithm therefore proceeds as follows:

1. Generate a random initial population uniformly in the input space;
2. $\theta^{(0)} \leftarrow$ mean f value on this population; $t \leftarrow 0$;
3. **repeat:**
4.   Update the density estimator model $p^{\theta^{(t)}}$;
5.   Generate more samples from the model $p^{\theta^{(t)}}$;
6. $\theta^{(t+1)} \leftarrow N$-th percentile of the data;
7.   retain only the points less than $\theta^{(t+1)}$; $t \leftarrow t+1$;

What is left is the choice of a parametric model so that it can be used and updated within acceptable CPU times and without requiring an excessive number of sample points. In [49] one first approximates the true distribution $p(X)$ with a distribution $\hat{p}(X)$ chosen from a reduced set of possibilities. The closest $\hat{p}(X)$ is defined as the one minimizing the Kullback-Leibler divergence $D(p\|\hat{p})$, a widely used measurement of the agreement between distributions. Finally, because identifying the optimal $\hat{p}(X)$ requires excessive computational resources, one resorts to a greedy construction of the distribution $\hat{p}(X)$.

Now that the path is clear, let us see the details. The joint probability distribution

$$p(X) = p(X_1|X_2,\ldots,X_n)p(X_2|X_3,\ldots,X_n)\ldots p(X_{n-1}|X_n)p(X_n)$$

is approximated by starting from pairwise conditional probabilities $p(X_i|X_j)$. In detail, one considers a class of probability distributions $\hat{p}_\pi(X)$, where each member is defined by a permutation π of the indices between 1 and n.

$$\hat{p}_\pi(X) = p(X_{\pi_1}|X_{\pi_2})p(X_{\pi_2}|X_{\pi_3})\cdots p(X_{\pi_{n-1}}|X_{\pi_n})p(X_{\pi_n}).$$

One aims at picking the permutation π such that $\hat{p}_\pi(X)$ has minimum divergence with the actual distribution. The Kullback-Leibler divergence $D(p\|\hat{p})$ to minimize is:

$$\begin{aligned} D(p\|\hat{p}_\pi) &= \int_{\mathscr{X}} p(X)\bigl(\log p(X) - \log \hat{p}_\pi(X)\bigr)\mathrm{d}X \\ &= E_p[\log p] - E_p[\log \hat{p}_\pi] \\ &= -h(p) + \underbrace{h(X_{i_1}|X_{i_2}) + h(X_{i_2}|X_{i_3}) + \cdots + h(X_{i_{n-1}}|X_{i_n}) + h(X_{i_n})}, \end{aligned} \quad (6.1)$$

where the properties of the logarithm and the definition of entropy h were used. Actually, the first term in the summation does not depend on π so that the cost function to be minimized is given by the rest of the summation (the terms within the underbrace). Because searching among all $n!$ permutations is too demanding, one adopts a greedy algorithm to fix indices sequentially, starting from index i_n, and fixing each index by minimizing the appropriate quantity as follows:

1. $i_n \leftarrow \arg\min_j \hat{h}(X_j)$;
2. **for** $k = n-1, n-2, \ldots, 1$:
3. $\quad i_k \leftarrow \arg\min_{j \notin \{i_{k+1},\ldots,i_n\}} \hat{h}(X_j|X_{i_{k+1}})$.

where $\hat{h}$ is the estimated empirical entropy. After the distribution is chosen, one generates samples as follows:

1. Randomly pick a value for X_{i_n} based on the estimated $\hat{p}(X_{i_n})$;
2. **for** $k = n-1, n-2, \ldots, 1$:
3. $\quad$ pick a value for X_{i_k} based on the estimated $\hat{p}(X_{i_k}|X_{i_{k+1}})$.

In the above technique, an approximated probability distribution is therefore found in a heuristic greedy manner. An alternative standard technique is that of *stochastic gradient ascent*. For simplicity, assume that f is positive, otherwise make it positive by applying a suitable transformation. One starts from an initial value for the parameters θ of the probabilistic construction model and does steepest descent where the exact gradient is estimated by sampling [216]. In detail, let us assume that the probabilistic sample generation model is determined by a vector of parameters $\theta \in \Theta$, and that the related probability distribution over generated solutions p_θ is differentiable. The combinatorial optimization problem is now substituted with a continuous optimization problem: Determine the model parameters leading to the highest expected value of f:

$$\theta^* = \arg\max_\theta E_{p_\theta} f(X)$$

Gradient ascent consists of starting from an initial $\theta^{(0)}$ and, at each step, calculating the gradient and updating $\theta^{(t+1)} = \theta^{(t)} + \varepsilon^{(t)} \nabla E(\theta^{(t)})$, where $\varepsilon^{(t)}$ is a suitably small step-size. Next, because

$$\nabla E(\theta) = \nabla \sum_X f(X) P_\theta(X) = \sum_X f(X) \nabla P_\theta(X)$$
$$= \sum_X f(X) P_\theta \frac{\nabla P_\theta(X)}{P_\theta} = \sum_X f(X) P_\theta \nabla \ln P_\theta, \qquad (6.2)$$

the gradient of the expectation can be substituted with an empirical mean from samples s extracted from the distribution S_t, obtaining the following update rule:

$$\theta^{(t+1)} = \theta^{(t)} + \varepsilon^{(t)} \sum_{s \in S_t} f(X) \nabla \ln P_{\theta^{(t)}}. \qquad (6.3)$$

We leave to the reader the gradient calculation, because it is dependent on the specific model.

More recent work considering possible dependencies is presented in [272, 273]. According to the authors, in designing practical population-based algorithms, it is sufficient to consider some selected crucial dependence relationships in terms of convergence. The difficulties in estimating the distribution from a finite-size population are also underlined as a challenging issue for discrete optimization problems. A priori knowledge to create probabilistic models for optimization is considered in [12]. *Prior* distributions related to the knowledge of parameter dependencies are used to create more accurate models for sample generation and for initializing starting points.

6.4 The Cross-Entropy Model

The *cross-entropy* method of [46, 220] builds upon the above ideas and upon [219], which considers an adaptive algorithm for estimating probabilities of rare events in stochastic systems. One starts from a distribution p_0, which is progressively updated to increase the probability of generating good solutions. If we multiply at each iteration the starting distribution by the function evaluation $\hat{p}(X) \propto p_t(X) f(X)$, we obtain higher probability values for higher function values. It looks like we are on the correct road: When n goes to infinity $p_n(X) \propto p_t(X)(f(X))^n$ the probability tends to be different from zero only at the global optima! Unfortunately, there is an obstacle: If p_t is a distribution that can be obtained by fixing parameters θ in our model, there is no guarantee that also $\hat{p}$ is going to be a member of this parametric family. Here our cross-entropy measure (CE) comes to the rescue; we will *project* $\hat{p}$ to the closest distribution in the parametric family, where closeness is measured by the Kullback-Leibler divergence:

$$D(\hat{p}\|p) = \sum_X \hat{p} \ln \frac{\hat{p}(X)}{p(X)} \qquad (6.4)$$

or, taking into account that $\sum_X \hat{p} \ln \hat{p}$ is constant when minimizing over p, by the *cross-entropy*:

$$h(\hat{p}|p) = -\sum_X \hat{p} \ln p(X). \tag{6.5}$$

Because our $\hat{p}$ is proportional to $p_t \times f$, the final distribution update is given by solving the maximization problem:

$$p_{t+1} = \arg\max_\theta \sum_X p_t(X) f(X) \ln p(X). \tag{6.6}$$

What holds for the original f also holds for monotonic transformations of f, which may also depend on memory, so that different functions can be used at different steps. An example is an indicator function $I(f < \theta^{(t)})$ as used in the MIMIC technique, or a Boltzmann function $f_T = \exp -f/T$, where the temperature T can be adapted. Large T values tend to smooth f differences leading to a less-aggressive search.

Similar to what was done for stochastic gradient descent, a fast sample approximation can substitute the summation over the entire solution space:

$$p_{t+1} = \arg\max_\theta \sum_{s \in S_t} f(X) \ln p(X). \tag{6.7}$$

We already anticipate the final part of the story: The above maximization problem usually cannot be solved exactly but again one can resort to stochastic gradient ascent, for example, starting from the previous p_t solution, and, again, remembering the pitfalls of local minima. If instead of a complete ascent one follows only one step along the estimated gradient, one actually recovers the stochastic gradient ascent rule:

$$\theta^{(t+1)} = \theta^{(t)} + \varepsilon^{(t)} \sum_{s \in S_t} f(s) \nabla \ln p_{\theta^{(t)}}(s). \tag{6.8}$$

The cross-entropy method therefore appears as a generalization of the SGA technique, allowing for time-dependent quality functions.

The theoretical framework above is surely of interest, unfortunately many issues have to be specified before obtaining an effective algorithm for a specific problem. In particular, the model has to be sufficiently simple to allow fast computation, but also sufficiently flexible to accommodate the structural properties of specific landscapes. In addition, local minima are haunting along the way of gradient ascent in the solution of the embedded optimization tasks implicit in the technique. An intrinsic danger is related to the "intensification" (exploitation) flavor of the approach: One tends to search where previous searches have been successful. Unfortunately, a new gold mine will never be found if the first iterations concentrate the exploration too much; see also Fig. 6.1. Contrary to gold mining, the reward in optimization is not for revisiting old good solutions but for rapidly exploiting a promising region and then moving on to explore uncharted territory.

The Greedy Randomized Adaptive Search (GRASP) framework [94] is based on repetitions of randomized greedy constructions and local search starting from the just constructed solution. Different starting points for local search are obtained by either stochastically breaking ties during the selection of the next solution component or by *relaxing the greediness*, i.e., putting at each greedy iteration a fraction of the most promising solution components in a restricted candidate list, from which the winner is randomly extracted. The bigger the candidate list, the larger the exploration characteristics of the construction. In spite of the name, there is actually no adaptation to previous runs in GRASP, but the opportunity arises for a set of self-tuning possibilities. For example, the size of the candidate list can be self-tuned, or statistics about relationships between final quality and individual components (or more complex structural properties learned from the previous constructions) can influence the choice of the next solution component to add, going toward the model-based search context. Preliminary investigations about Reactive-GRASP are presented in [114, 212]. The purpose is to adapt the size of the candidate list by considering information gathered about the quality of solutions generated with different values. In detail, a parameter α is defined as the fraction of elements that are considered in the restricted candidate list; a set of possible values $\{\alpha_1, \ldots, \alpha_n\}$ is considered. A value of α is chosen at each construction with probability p_i. The adaptation acts on the probabilities. At the beginning they are uniform, $p_i = 1/n$. After a number of GRASP constructions, the average solution value a_i obtained when using α_i is computed and the probabilities are updated so that they become proportional to these average values (of course scaled so that they sum up to one). The power of reactive-GRASP is derived both from the additional diversification implicit when considering different values for α and from the reactive adaptation of the probabilities.

6.5 Adaptive Solution Construction with Ant Colonies

In spite of the practical challenges involved in defining and using models, some approaches based on analogies with ant colonies behavior [86] are based on similar principles: Solution construction happens via a model that is influenced by the quality of previously generated solutions. Like it is the case for genetic algorithms, these algorithms are characterized by the use of sexy terms to describe the relevant parameters, in this case inspired by the behavior of biological ants. One of these terms is *pheromone*, a chemical that triggers a natural behavioral response in another member of the same species, used as a cute name for parameters affecting the probabilities of choosing the next solution component during the greedy construction.

It is of interest that the most successful ACO applications are in fact combinations of the basic technique with advanced memory-based features, local search, and problem-specific heuristic information about the a priori desirability of the different solution components. Figure 6.4 shows a generic ACO framework: The main feature

Object	Scope	Meaning
f	(input)	Function to minimize
n_{ants}	(input)	Number of ants
x_{best}	(output)	Best configuration found
T	(internal)	Pheromone trail parameter set
x	(internal)	Current configuration
f_{best}	(internal)	Best objective value
G_{iter}	(internal)	Set of solutions found by ants in current iteration
INITPHEROMONE	(function)	Set initial trail values to predefined constant
CONSTRUCTSOLUTION(T)	(function)	Build probabilistically a solution based on current pheromone values T
VALIDSOLUTION(x)	(function)	Check whether a solution is legal
LOCALSEARCH(x)	(function)	Optionally apply a minimization step to the current configuration
PHEROMONEUPDATE(...)	(function)	Compute new pheromone parameters

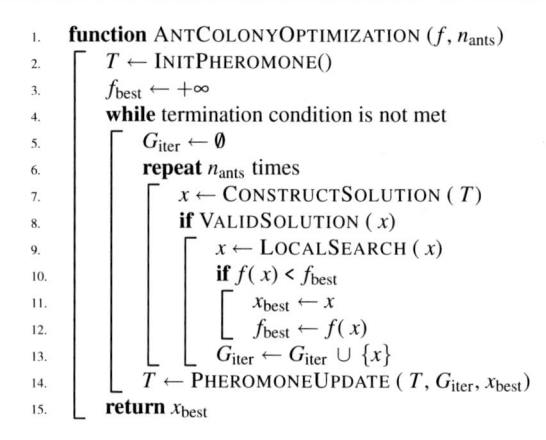

```
 1. function ANTCOLONYOPTIMIZATION ($f$, $n_{ants}$)
 2.   $T \leftarrow$ INITPHEROMONE()
 3.   $f_{best} \leftarrow +\infty$
 4.   while termination condition is not met
 5.     $G_{iter} \leftarrow \emptyset$
 6.     repeat $n_{ants}$ times
 7.       $x \leftarrow$ CONSTRUCTSOLUTION ($T$)
 8.       if VALIDSOLUTION ($x$)
 9.         $x \leftarrow$ LOCALSEARCH ($x$)
10.         if $f(x) < f_{best}$
11.           $x_{best} \leftarrow x$
12.           $f_{best} \leftarrow f(x)$
13.         $G_{iter} \leftarrow G_{iter} \cup \{x\}$
14.     $T \leftarrow$ PHEROMONEUPDATE ($T$, $G_{iter}$, $x_{best}$)
15.   return $x_{best}$
```

Fig. 6.4 Outline of a generic Ant Colony Optimization framework

of the algorithm is the maintenance of a vector T of "pheromone" parameters that are used to drive probabilistically the generation of new configurations (at each iteration, n_{ants} ants generate one such configuration each). The vector T is updated at the end of an iteration on the basis of the valid configurations found by every ant and the best solution so far. The max-min ant-system algorithm of [245] improves some of the weak points of the original algorithm by achieving a strong exploitation of the search history (only the best solutions are allowed to add pheromone) and by limiting the strength of the pheromone's trials to effectively avoid premature convergence caused be excessive intensification.

6.6 Modeling Surfaces for Continuous Optimization

As expected, continuous optimization shares the objective of using knowledge derived from a set of evaluations of $f(X)$ in order to build *models* of the f surface, models intended to identify the most promising regions to explore.

The issue of using knowledge from samples to build models is deeply related to statistics and machine learning. In fact, an extensive presentation of supervised and reinforcement-based machine learning techniques will be presented later in Chaps. 7 and 8. In this section, we present some motivations and some history related to the *response surfaces*, *experimental design*, and *kriging* techniques.

An important motivation to consider modeling is that evaluating the objective function f is in some cases terribly expensive. Dr. Krige, a South African mining engineer and originator of the *kriging* method, was in charge of evaluating mineral resources for gold-mining companies. A single evaluation consisted for him in excavating a new mine, quite a slow and expensive process! It is natural that he thought about ways to reduce the number of f evaluations by predictive models aimed at using in the best possible manner the data related to previous mining experiments. The same situation is present when the evaluation of f has to do with running an experiment, or running an expensive computer simulation. For example, f can be the yield of a complex manufacturing process, without an analytical closed-form expression, so that evaluation requires running the plant for some days, spending human and material resources.

Response surfaces are popular techniques to develop fast surrogates for time-consuming computer simulations. A recent clear taxonomy of global optimization methods based on response surfaces is presented in [153]. The *surface* $\hat{f}$ in the method is built by starting from a (small) set of n sample points $\{(X_1, f(X_1)), \ldots, (X_n, f(X_n))\}$, where f is a possibly stochastic function, and by deriving a parametric "interpolating" surface. Typically, the surface is intended to model the expected value $E[f(X)]$ at point X. After the surface is known, one can devise strategies for placing the next point X_{n+1} to be tested through an experiment or a computer simulation in order to derive $f(X_{n+1})$. A related area is that of *experimental design* based on statistics. *Design of experiments* includes the methodological design of all information-gathering processes involved in a scientific experiment, including strategies to place the next experimental setting, and to determine the relevance of the different input variables on the output value $f(X)$.

Nontrivial choices are involved in response surface techniques and counterexamples where a specific technique fails in the presence of sets of deceptive sample points are easy to demonstrate. First, one must select a model that is appropriate to the optimization instance. For example: Is the model based on interpolation (the surface is forced to pass through all sample $f(X_i)$ values) or on minimizing some form of average error (e.g., the sum of squared errors from some predetermined functional form)? What is the functional form to be fitted to the data points? In some cases, a linear combination of basis functions is used; see also the discussion of support vector machine (SVM) methods in Chap. 7. In this case, the basis functions can be fixed or have parameters that can be tuned (as in kriging). The choice of model is related

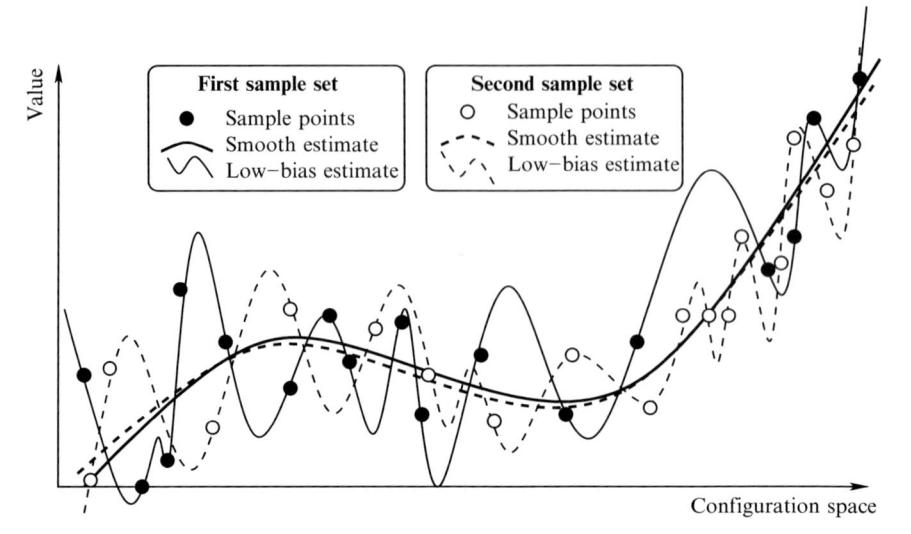

Fig. 6.5 Bias–variance tradeoff in machine learning and modeling: Two sample sets from the same experiment can give markedly different models if one aims at minimizing the error, while approximations using less parameters are more stable, even if the error on the examples is larger

to the well-known *bias–variance tradeoff* in machine learning (Fig. 6.5). The *bias* term measures the difference between the true f and the average of the predicted values at X. The *variance* term measures how sensitive the predicted value at X is to random fluctuations in the dataset. If an analyst wants to reduce variance, he expects an increase in bias. If an analyst wants to reduce bias, he expects an increase in variance. The high-school example is that of least-squares fitting with a polynomial. If the degree of the polynomial is less than the number of sample points, one will obtain a smooth curve that is rather stable for small variations in the data points. If the number of parameter increases, the interpolation error will decrease (to zero as soon as the degree is equal to the number of data points) but wild and unstable oscillations will be present between the sample points, critically dependent on the detailed sample point values.

Even if the model is appropriate for the optimization task, its usage is not trivial. For example, is the strategy of positioning the next experimental point in correspondence to the minimizer of $\hat{f}$ the best one? Not necessarily, in particular, if an estimate of the expected *error in the interpolator* is available. It is reasonable that the error will be small in areas where many sample points X_i are present, much larger in areas very far from the sample points. Here one faces the *exploration–exploitation dilemma*: Is it better to concentrate the search in the promising and well-known regions (the ones with small error estimate) or to explore the less-known areas, where radically better solutions may be present? In many cases a properly balanced strategy is followed, by exploring areas where the expected $\hat{f}$ value minus a suitable estimated error (the standard error) is minimum.

6.6 Modeling Surfaces for Continuous Optimization

This section is intended only as an appetizer for the additional details in Chap. 7 about supervised learning (response surface is just an historical way to denote a learnt model), and so let us mention only the widely used *kriging* model. The predictor is a combination of polynomial interpolation and *basin function* terms. In detail, the predicted value at point Y is given by:

$$\hat{f}(Y) = \sum_{k=1}^{m} a_k \pi_k(X) + \sum_{j=1}^{n} b_j \phi(Y - X_i), \qquad (6.9)$$

where $\{\pi_k(X)\}$ is a basis of the set of all polynomials in X of a given maximum degree, and $\phi(Z)$ are basis functions that are then centered on the X_i sample points. One of the widely used basis function for *kriging* has a smoothly decaying form as:

$$\phi(Z) = \exp - \sum_{l=1}^{d} \theta_l \|z_l\|^{p_l}, \qquad (6.10)$$

where the parameters satisfy $\theta_l \geq 0$ and $0 < p_l \leq 2$. The parameters take care of different sensitivities along different coordinates (different scales possibly related to different units of measures) and different decay rates.

An approach to noisy and costly optimization based on nonparametric reasoning about simple geometric relationships is developed in [7].

Chapter 7
Supervised Learning

> The task of the excellent teacher is to stimulate "apparently ordinary" people to unusual effort. The tough problem is not in identifying winners: it is in making winners out of ordinary people.
> (K. Patricia Cross)

> A mind is a fire to be kindled, not a vessel to be filled.
> (Plutarch)

7.1 Learning to Optimize, from Examples

Small children learn to read and write through a teacher who presents on the blackboard examples of the letters of the alphabet while uttering the corresponding sounds. The pupils learn in a gradual manner after repeated presentations of examples, by training their biological neural networks how to associate images to symbols and sounds.

Supervised learning means learning from examples, sets of associations between inputs and outputs. In detail, one aims at building an association $y = \hat{f}(x)$ between input x (in general a vector of parameters called *features*) and output y, by starting from a set of labeled examples. An input example can be the image of a letter of the alphabet, with the output corresponding to the letter symbol. In classification, i.e., recognition of the class of a specific object described by features x, the output is a suitable code for the class. To take care of indecision, e.g., consider an image where one is undecided between the letter 0 (zero) and the letter O (like in Old), one can have a real-valued output ranging between zero and one interpreted as the posterior probability for a given class given the input values.

A supervised learning process has to do with selecting a flexible model $\hat{f}(x;w)$, where the flexibility is given by some tunable parameters – or weights – w, and with selecting a process to determine the best value w^* of the parameter in a manner depending on the examples.

The learned model has to work correctly on the examples in the *training set*, and a reasonable criterion seems to require that the error between the correct answer (given by the example label) and the outcome predicted by the model is minimized. Supervised learning therefore becomes *minimization of a specific error function*, depending on parameters w. If the function is differentiable, a naïve classroom approach consists of using gradient descent, also known as steepest descent. One iterates by calculating the gradient of the function with respect to the weights and by taking a small step in the direction of the negative gradient. This is in fact the popular technique in neural networks known as learning by *backpropagation* of the error. To continue with the analogy between supervised learning and the classroom, gradually minimizing the error function can be seen as the teacher correcting the children in their first attempts of reading. Now, minimization of an error function is indeed a first critical component, but not the only one. The *model complexity*, related to the number of parameters and to the modeling flexibility, is also critical. If the complexity is too large, learning the examples with zero errors becomes trivial, but the problems begin when new data is presented as input. In the human metaphor, if learning becomes rigid memorization of the examples without grasping the underlying model, pupils will have difficulties in generalizing to new cases. This is related to the *bias–variance* dilemma mentioned in Chap. 6 and requires care in model selection, or minimization of a weighted combination of model error plus model complexity. A detailed presentation of first- and second-order neural network learning techniques is given in [18], while a comprehensive treatment of pattern classification is present in [89].

In this chapter we present a summary of the main techniques used in supervised learning, derived from [56], and some example applications in the field of performance prediction, algorithm selection, and automated parameter tuning.

7.2 Techniques

To fix the notation, a training set of ℓ tuples is considered, where each tuple is of the form (x_i, y_i), $i = 1, \ldots, \ell$, $x_i \in \mathbb{R}^d$ being a vector of input parameter values and y_i being the measured outcome to be learned by the algorithm. We consider two possible problems:

- *regression problems*, where y_i is a real value; for example, if we want to predict the length of an optimization run, x_i contains the parameter values of the ith run of an optimization algorithm, consisting of features extracted from the instance, while y_i is its measured duration.
- *classification problems*, where y_i belongs to a finite set, e.g., $y_i = \pm 1$ or $y_i \in \{1, \ldots, N\}$. For instance, given some features of the search space, we want to determine the best algorithm out of a small set.

7.2.1 Linear Regression

A linear dependence of the output from the input features usually is the first model to be tested. If you remember calculus, every smooth (differentiable) function can be approximated around an operating point x_c with its Taylor series approximation. The second term of the series is linear, given by a scalar product between the gradient $\nabla f(x_c)$ and the displacement vector, the additional terms go to zero in a quadratic manner:

7.2 Techniques

$$f(x) = f(x_c) + \nabla f(x_c) \cdot (x - x_c) + O(\|x - x_c\|^2). \tag{7.1}$$

Therefore, if the operating point of the smooth systems is close to a specific point x_c, a linear approximation is a reasonable place to start.

As one expects, the model is simple, it can be easily trained, and the computed weights provide a direct explanation of the importance of the various features. Therefore, do not complicate models unless this is required by your application. Let us see the details.

The hypothesis of a linear dependence of the outcomes on the input parameters can be expressed as

$$y_i = w^T \cdot x_i + \varepsilon_i,$$

where $w = (w_1, \ldots, w_d)$ is a (column) vector of *weights* to be determined and ε_i is the error, which is normally assumed to have Gaussian distribution. In other words, we are looking for the weight vector w so that the linear function

$$\hat{f}(x) = w^T \cdot x \tag{7.2}$$

approximates as closely as possible our experimental data. We can achieve this goal by finding the vector w^* that minimizes the sum of the squared errors:

$$\text{ERROR}(w) = \sum_{i=1}^{\ell} (w^T \cdot x_i - y_i)^2. \tag{7.3}$$

Standard techniques for solving the *least-squares* problem, by equating the gradient to zero, lead to the following optimal value for w:

$$w^* = (X^T X)^{-1} X^T y, \tag{7.4}$$

where $y = (y_1, \ldots, y_\ell)$ and X is the matrix whose rows are the x_i vectors.

The matrix $(X^T X)^{-1} X^T$ is called *pseudo-inverse* and it is a natural generalization of a matrix inverse to the case in which the matrix is nonsquare. If the matrix is invertible and the problem can be solved with zero error, one goes back to the inverse, but in general, for example if the number of examples is larger than the number of weights, aiming at a *least-square* solution avoids the embarrassment of not having an exact solution and provides a statistically sound "compromise" solution. In the real world, exact models are not compatible with the noisy characteristics of nature and of physical measurements and it is not surprising that *least-square* and *pseudo-inverse* beasts are among the most popular tools.

Let us note that the minimization of squared errors has a physical analogy to the spring model presented in Fig. 7.1. Imagine that every sample point is connected by a vertical spring to a rigid bar, the physical realization of the best fit line. All springs have equal elastic constants and zero extension at rest. In this case, the potential energy of each spring is proportional to the square of its length, so that (7.3) describes the overall potential energy of the system up to a multiplicative constant. If one starts and lets the physical system oscillate until equilibrium is reached, with some friction to damp the oscillations, the final position of the rigid bar can be read out to obtain the least-square fit parameters; an analog computer for line fitting!

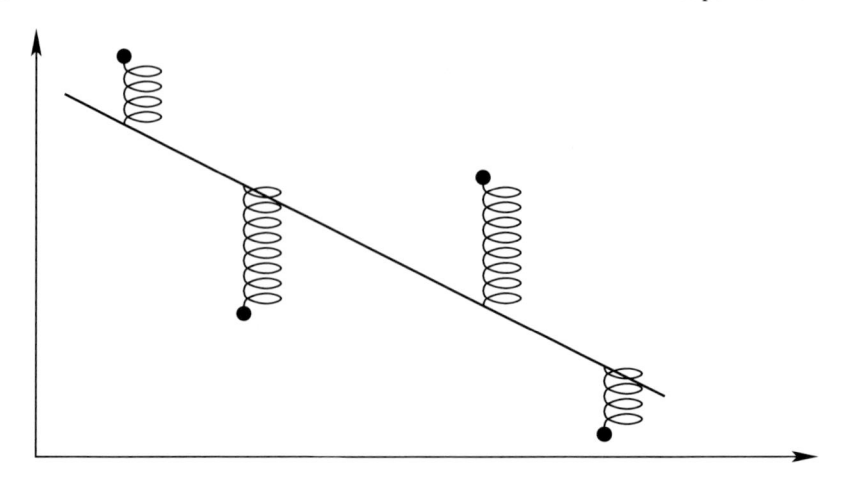

Fig. 7.1 Physics gives an intuitive spring analogy for least-squares fits. The best fit is the line that minimizes the overall potential energy of the system (proportional to the sum of the squares of the spring length)

7.2.1.1 Dealing with Nonlinear Dependencies

A function in the form $f(x) = w^T x$ is too restrictive to be useful in most cases. In particular, it assumes that $f(0) = 0$. It is possible to change from a *linear* to an *affine* model by inserting a constant term: $f(x) = w_0 + w^T \cdot x$. The constant term can be incorporated into the dot product, by defining $x = (1, x_1, \ldots, x_d)$, so that (7.2) remains valid.

The insertion of a constant term is a special case of a more general technique aiming at the definition of nonlinear dependencies while remaining in the (easier) context of linear least-squares approximations. This apparent contradiction is solved by a trick; the model remains linear and it is applied to *nonlinear features* calculated from the raw input data instead of the original input x. In other words, one defines a set of functions

$$\phi_1, \ldots, \phi_n : \mathbb{R}^d \to \mathbb{R}$$

that map the parameters space into some more complex space, in order to apply the linear regression to the vector $\phi(x) = (\phi_1(x), \ldots, \phi_n(x))$ rather than to x directly.

For example, if $d = 2$ and $x = (x_1, x_2)$ is a parameter vector, a quadratic dependence of the outcome can be obtained by defining the following *basis functions*:

$$\phi_1(x) = 1, \quad \phi_2(x) = x_1, \quad \phi_3(x) = x_2, \quad \phi_4(x) = x_1 x_2, \quad \phi_5(x) = x_1^2, \quad \phi_6(x) = x_2^2.$$

Note that $\phi_1(x)$ is defined in order to allow for a constant term in the dependence. The linear regression technique described above is then applied to the six-dimensional vectors obtained by applying the basis functions, and not to the original two-dimensional parameter vector.

More precisely, one looks for a dependence given by a scalar product between a vector of weights w and a vector of features $\phi(x)$, as follows:

7.2 Techniques

$$\hat{f}(x) = w^T \cdot \phi(x).$$

In other words, the output is a weighted sum of the features. Let $x'_i = \phi(x_i)$, $i = 1,\ldots,\ell$, be the transformations of the training input tuples x_i. If X' is the matrix whose rows are the x'_i vectors, then the optimal weights with respect to the least-squares approximation are computed as above:

$$w^* = (X'^T X')^{-1} X'^T y. \tag{7.5}$$

7.2.1.2 Preventing Numerical Instabilities

As the reader may notice, when the number of examples is large, (7.5) is the solution of a linear system in the over-determined case (more linear equations than variables). In particular, matrix $X^T X$ must be nonsingular, and this can only happen if the training set points $x_1,\ldots,x_\ell$ do not lie in a proper subspace of $\mathbb{R}^d$, i.e., they are not "aligned." In many cases, even though $X^T X$ *is* invertible, the distribution of the training points is not generic enough to make it *stable*. *Stability* here means that small perturbations of the sample points lead to small changes in the results. An example is given in Fig. 7.2, where a bad choice of sample points (in the bottom plot, x_1 and x_2 are not independent) makes the system much more dependent on noise, or even to rounding errors.

If there is no way to modify the choice of the training points, the standard mathematical tool to ensure numerical stability when sample points cannot be distributed at will is known as *ridge regression*. It consists of the addition of a *regularization* term to the (least-squares) error function to be minimized:

$$\mathrm{ERROR}(w;\lambda) = \sum_{i=1}^{\ell} (w^T \cdot x_i - y_i)^2 + \lambda w^T \cdot w.$$

The minimization with respect to w leads to the following:

$$w^* = (\lambda I + X^T X)^{-1} X^T y.$$

Intuitively, the insertion of a small diagonal term makes the inversion more robust. Moreover, one is actually asking that the solution takes the size of the weight vector into account, to avoid steep interpolating planes such as the one in the bottom plot of Fig. 7.2.

If you are interested, the theory justifying the approach is based on *Tichonov regularization*, which is the most commonly used method for curing *ill-posed* problems. A problem is ill-posed if no unique solution exists because there is not enough information specified in the problem, for example because the number of examples is limited. It is necessary to supply extra information or assumptions.

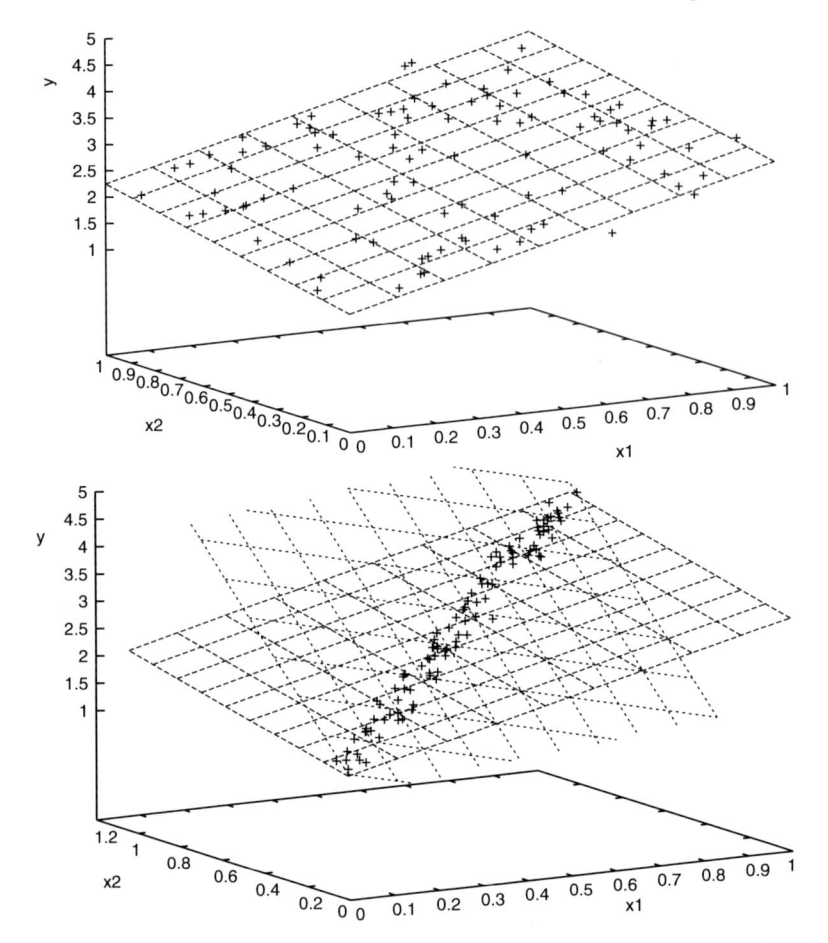

Fig. 7.2 A well-spread training set *(top)* provides a stable numerical model, whereas a bad choice of sample points *(bottom)* may result in wildly changing and unpredictable results

7.2.2 Bayesian Locally Weighted Regression

In the previous section we have seen how to determine the coefficients of a linear dependence, and how to extend the technique to allow for nonlinear relationships by encoding the nonlinear part in a base function vector. The approach requires that the overall form of the dependence is stated by the researcher ("let's try a cubic function here!"), but in many cases, this is impossible because the dependence is too complex. If one visualizes y as the elevation of a terrain, a mountain area presents too many hills, peaks, and valleys to be modeled by simple nonlinear terms such as products of the basic variables.

The approach described in this section is based upon the following idea: A relationship based on experimental data can be safely assumed to be "locally" linear, although the overall global dependence can be quite complex. If the model must be evaluated at different points, then we can still use linear regression, provided that training points *near* the evaluation point are considered "more important" than distant ones. We encounter a very general principle here: In learning (natural or automated), similar cases are usually deemed more relevant than very distant ones. For example, case-based reasoning solves new problems based on the solutions of similar past problems. A lawyer who advocates a particular outcome in a trial based on similar legal precedents is using case-based reasoning, but let us get back to the math now.

Locally weighted regression [71] is characterized as a *lazy* memory-based technique, meaning that all points and evaluations are stored and a specific model is built *on-demand* only when a specified query point demands an output. The occasional lack of sample points near the query point can pose problems in estimating the regression coefficients with a simple linear model. Hence *Bayesian* locally weighted regression [9], denoted as B-LWR, is used if one can specify prior information about what values the coefficients should have when there is not enough data to determine them. The usual power of Bayesian techniques derives from the *explicit* specification of the modeling assumptions and parameters (for example, a *prior distribution* can model our initial knowledge about the function) and the possibility to model not only the expected values but entire probability distributions. For example *confidence intervals* can be derived to quantify the uncertainty in the expected values.

To predict the outcome of an evaluation at a point q (named a *query point*), linear regression is applied to the training points. To enforce locality in the determination of the regression parameters (near points are more relevant), each sample point is assigned a *weight* that decreases with its distance from the query point. Note that, in the neural networks community, the term *weight* refers to the parameters of the model being computed by the training algorithm, while, in this case, it measures of the importance of each training sample. To avoid confusion we use the term *significance* and the symbol s_i (and S for the diagonal matrix used below) for this different purpose.

In the following we shall assume, as explained in Sect. 7.2.1, that a constant 1 is appended to all input vectors x_i to include a constant term in the regression, so that the dimensionality of all equations is actually $d+1$.

The weighted version of least-squares fit aims at minimizing the following weighted error (compare with (7.3), where weights are implicitly uniform):

$$\text{ERROR}(w; s_1, \ldots, s_n) = \sum_{i=1}^{\ell} s_i (w^\text{T} \cdot x_i - y_i)^2. \tag{7.6}$$

From the viewpoint of the spring analogy discussed in Sect. 7.2.1, the distribution of different weights to sample points corresponds to using springs with a different elastic constant (strength), as shown in Fig. 7.3. Minimization of (7.6) is obtained by requiring its gradient with respect to w to be equal to zero, and we obtain the following solution:

$$w^* = (X^\text{T} S^2 X)^{-1} X^\text{T} S^2 y, \tag{7.7}$$

where $S = \text{diag}(s_1, \ldots, s_d)$. Note that (7.7) reduces to (7.4) when all weights are equal.

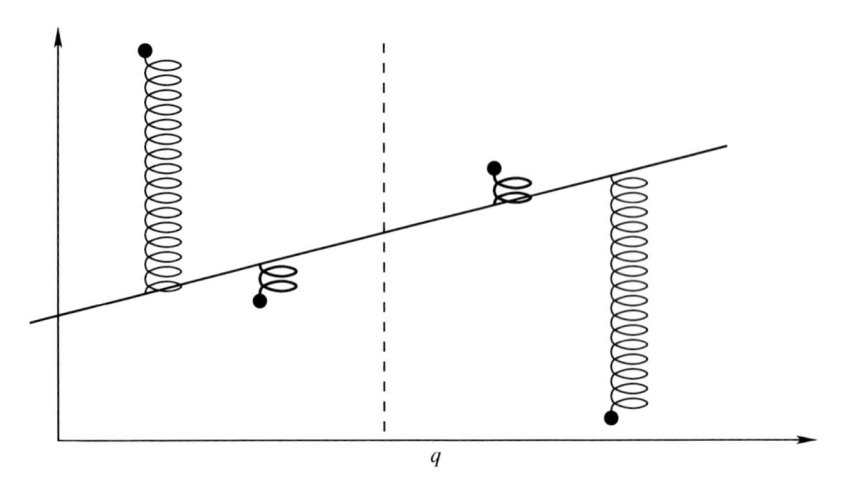

Fig. 7.3 The spring analogy for the weighted least-squares fit (compare with Fig. 7.1). Now springs have different elastic constants, thicker meaning harder, so that their contribution to the overall potential energy is weighted. In the above case, harder springs are for points closer to the query point q

While weighted linear regression is often used when samples are of heterogeneous nature (e.g., they suffer from different experimental errors), we are interested in assigning weights according to their distance from the query point. A common *kernel function* used to set the relationship between weight and distance is

$$s_i = \exp\left(-\frac{\|x_i - q\|^2}{K}\right),$$

where K is a parameter measuring the "kernel width," i.e., the sensitivity to distant sample points; if the distance is much larger than K, the significance rapidly goes to zero.

An example is given in Fig. 7.4 (top), where the model must be evaluated at query point q. Sample points x_i are plotted as circles, and their significance s_i decreases with the distance from q and is represented by the interior shade, black meaning highest significance. The linear fit (solid line) is computed by considering the significance of the various points and is evaluated at q to provide the model's value at that point. The significance of each sample point and the subsequent linear fit are recomputed for each query point, providing the curve shown in Fig. 7.4 (bottom).

Up to this point, no assumption has been made on the nature of coefficients to be determined. The prior assumption on the distribution of coefficients w, leading to Bayesian LWR, is that it is distributed according to a multivariate Gaussian with zero mean and covariance matrix Σ, and the prior assumption on σ is that $1/\sigma^2$ has a Gamma distribution with k and θ as the shape and scale parameters. Since one uses a weighted regression, each data point and the output response are weighted using

7.2 Techniques

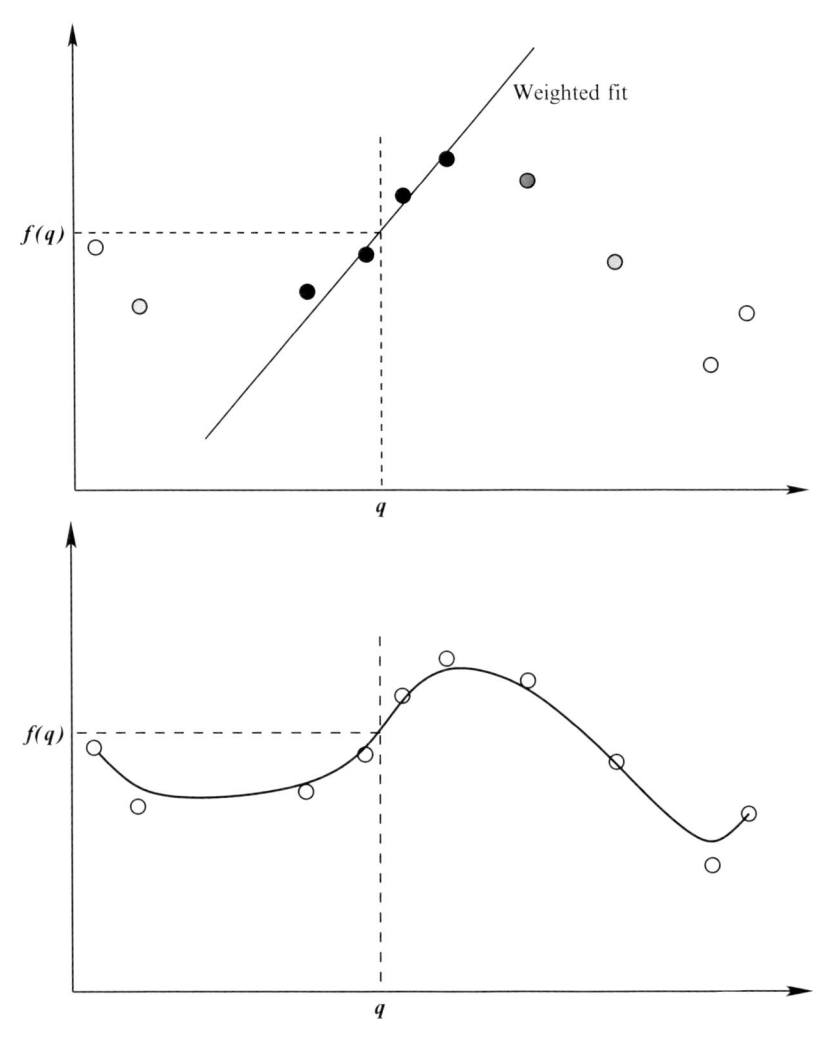

Fig. 7.4 *Top*: evaluation of LWR model at query point q, sample point significance is represented by the interior shade. *Bottom*: Evaluation over all points, each point requires a different linear fit

a Gaussian weighting function. In matrix form, the weights for the data points are described in $\ell \times \ell$ diagonal matrix $S = \mathrm{diag}(s_1, \ldots, s_\ell)$, while $\Sigma = \mathrm{diag}(\sigma_1, \ldots, \sigma_\ell)$ contains the prior variance for the w distribution.

The local model for the query point q is predicted by using the marginal posterior distribution of w whose mean is estimated as

$$\bar{w} = (\Sigma^{-1} + X^T S^2 X)^{-1} (X^T S^2 y). \tag{7.8}$$

Note that the removal of prior knowledge corresponds to having infinite variance on the prior assumptions, therefore Σ^{-1} becomes null and (7.8) reduces to (7.7). The matrix $\Sigma^{-1} + X^T S^2 X$ is the weighted covariance matrix, supplemented by the effect of the w priors. Let us denote its inverse by V_w. The variance of the Gaussian noise based on ℓ data points is estimated as

$$\sigma^2 = \frac{2\theta + (y^T - w^T X^T) S^2 y}{2k + \sum_{i=1}^{\ell} s_i^2}.$$

The estimated covariance matrix of the w distribution is then calculated as

$$\sigma^2 V_w = \frac{(2\theta + (y^T - w^T X^T) S^2 y)(\Sigma^{-1} + X^T S^2 X)}{2k + \sum_{i=1}^{\ell} s_i^2}.$$

The degrees of freedom are given by $k + \sum_{i=1}^{\ell} s_i^2$. Thus the predicted output response for the query point q is

$$\hat{y}(q) = q^T \bar{w},$$

while the variance of the mean predicted output is calculated as:

$$\text{Var}(\hat{y}(q)) = q^T V_w q \sigma^2. \tag{7.9}$$

7.2.3 Using Linear Functions for Classification

Section 7.2.1 considered a (linear) function that nearly matches the observed data, for example, by minimizing the sum of squared errors.

Some optimization issues, however, allow for a small set of possible outcomes (e.g., should we restart or not? Should we apply a local search neighborhood of radius one or two?). One is faced with a *classification* problem.

Let the outcome variable be two-valued (e.g., ± 1). In this case, linear functions can be used as discriminants, and the goal of the training procedure becomes that of finding the best coefficient vector w so that the classification procedure

$$y = \begin{cases} +1 & \text{if } w^T \cdot x \geq 0 \\ -1 & \text{otherwise} \end{cases} \tag{7.10}$$

performs the best classification. Depending on the problem, the criterion for deciding the "best" classification changes. Some possible criteria are:

- The number of classification errors is minimum
- The *precision* (ratio of correct classifications within positive samples) or the *recall* (ratio of correct outcomes within positive classifications) is maximum
- A weighted average of the above is optimized (e.g., the harmonic mean, also known as the F_1 index).

7.2 Techniques

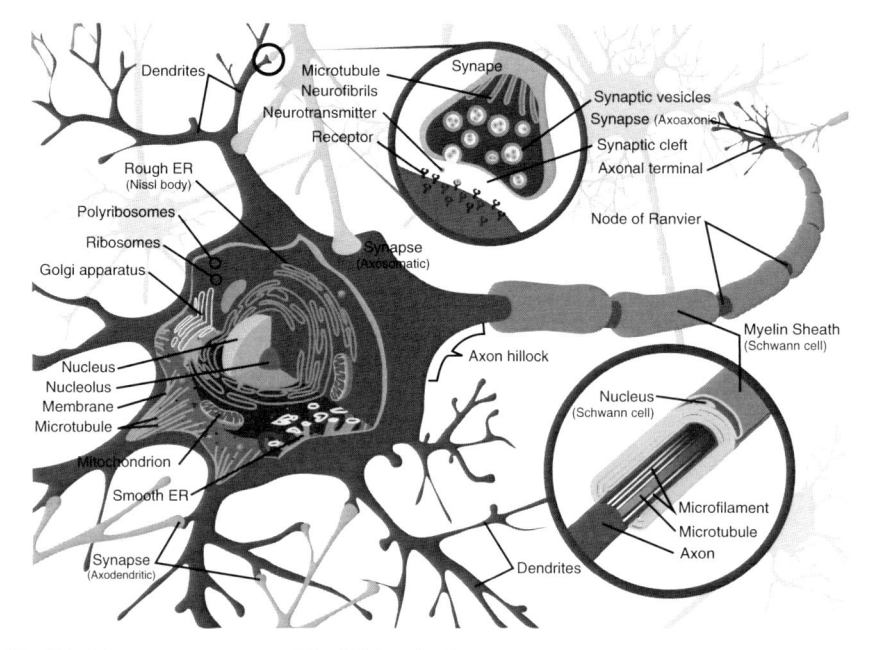

Fig. 7.5 The neuron (courtesy of the Wikimedia Commons)

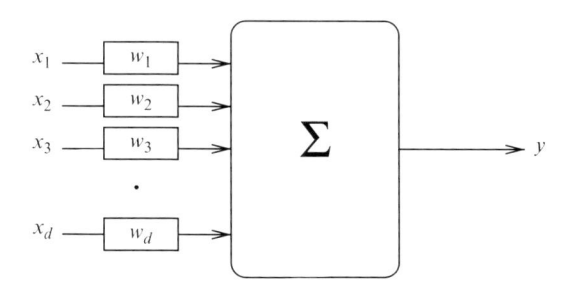

Fig. 7.6 The perceptron: The output is obtained by a weighted sum of the inputs passed through a final threshold function

Linear functions for classification have been known under many names, the historic one being *perceptron*, a name that stresses the analogy with biological neurons. Neurons (see Fig. 7.5) communicate via chemical synapses, in a process known as *synaptic transmission*. The fundamental process that triggers synaptic transmission is a propagating electrical signal that is generated by exploiting the electrically excitable membrane of the neuron. This signal is generated (the neuron output *fires*) if and only if the results of incoming signals combined with excitatory and inhibitory synapses and integrated surpasses a given threshold. Figure 7.6 therefore can be seen as the abstract and functional representation of a single nerve cell.

The algorithms for determining the best separating linear function (geometrically identifiable with a hyperplane) depend on the chosen classification criteria. Section 7.2.5 summarizes a sound theory for linear separation, extensible to the nonlinear case.

7.2.4 Multilayer Perceptrons

A multilayer perceptron (MLP) neural network is composed of a large number of highly interconnected units (*neurons*) working in parallel to solve a specific problem and organized in layers with a feed-forward information flow (no loops).

The architecture of the multilayer perceptron is organized as follows: The signals flow sequentially through the different layers from the input to the output layer. The intermediate layers are called *hidden* layers because they are not visible at the input or at the output. For each layer, each unit first calculates a scalar product between a vector of weights and the vector given by the outputs of the previous layer. A *transfer function* is then applied to the result to produce the input for the next layer Fig. 7.7. A popular smooth and *saturating* transfer function (the output saturates to zero for large negative signals, to one for large positive signals) for the hidden layers is the sigmoidal function:

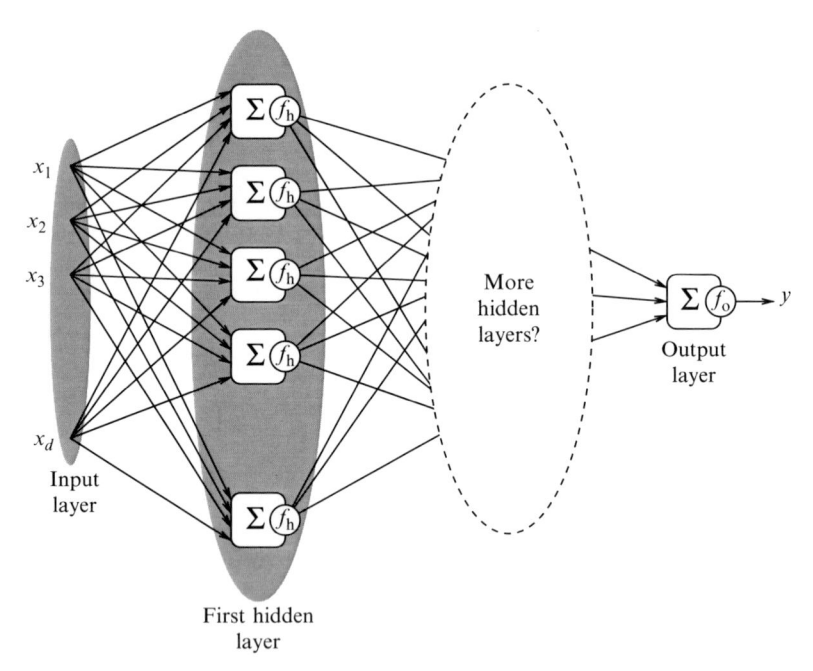

Fig. 7.7 Multilayer perceptron: The nonlinearities introduced by the transfer functions at the intermediate (hidden) layers permit arbitrary continuous mappings to be realized

$$f(x) = \frac{1}{1+e^{-x}}.$$

Other transfer functions can be used for the output layer; for example, the identity function can be used for unlimited output, while the sigmoidal function is more suitable for "yes/no" classification problems.

Many training techniques have been proposed for weight determination. An example is the one-step-secant (OSS) method with fast line searches [18] using second-order derivative information. The time complexity of the learning phase is usually high, although it varies greatly according to the heuristic adopted for calculating the weights (usually iterative) and to the halting condition. A large number of passes is required, each involving computation of the outcome for every training set point, followed by modification of weights.

On the other hand, calculating the outcome of an MLP neural network is a straightforward procedure. While perceptrons are limited in their modeling power to classification cases where the patterns of the two different classes can be separated by a hyperplane in input space, MLPs are universal approximators [137]: An MLP with one hidden layer can approximate any continuous function to any desired accuracy, subject to a sufficient number of hidden nodes.

7.2.5 Statistical Learning Theory and Support Vector Machines

We mentioned before that minimizing the error on a set of example is not the only objective of a statistically sound learning algorithm, also the modeling architecture has to be considered. Statistical learning theory [256] provides mathematical tools for *deriving unknown functional dependencies* on the basis of observations. A shift of paradigm occurred in statistics starting from the 1960s: In the previous paradigm based on Fisher's research in the 1920–1930s, in order to derive a functional dependency from observations one had to know the detailed form of the desired dependency and to determine only the values of a finite number of *parameters* from the experimental data. The new paradigm does not require the detailed knowledge, and proves that some general properties of the set of functions to which the unknown dependency belongs are sufficient to estimate the dependency from the data. *Non-parametric* technique is a term used for these flexible models, which can be used even if one does not know a detailed form of the input–output function. The MLP model described before is an example.

While we refer to [256] for a detailed presentation of the theory, a brief summary of the main methodological points of statistical learning theory is included to motivate the use of SVMs as a learning mechanism for intelligent optimization and reactive search applications; see [59, 203, 256] for more details.

Let $P(x,y)$ be the unknown probability distribution from which the examples are drawn. The learning task is to learn the mapping $x_i \to y_i$ by determining the values of the parameters of a function $f(x,w)$. The function $f(x,w)$ is called *hypothesis* and the the set $\{f(x,w), w \in \mathscr{W}\}$ is called the *hypothesis space* and denoted by $\mathscr{H}$.

$\mathscr{W}$ is the set of abstract parameters. A choice of the parameter $w \in \mathscr{W}$, based on the labeled examples, determines a "trained machine."

The *expected test error* or *expected risk* of a trained machine for the classification case is:

$$R(w) = \int \|y - f(x, w)\| \mathrm{d}P(x, y), \qquad (7.11)$$

while the *empirical risk* $R_{\text{emp}}(w)$ is the mean error rate measured on the training set:

$$R_{\text{emp}}(w) = \frac{1}{\ell} \sum_{i=1}^{\ell} \|y_i - f(x_i, w)\|. \qquad (7.12)$$

The classical learning method is based on the *empirical risk minimization* (ERM) inductive principle: One approximates the function $f(x, w^*)$, which minimizes the risk in (7.11), with the function $f(x, \hat{w})$, which minimizes the empirical risk in (7.12).

The rationale for the ERM principle is that, if R_{emp} converges to R in probability (as guaranteed by the law of large numbers), the minimum of R_{emp} may converge to the minimum of R. If this does not hold, the ERM principle is said to be *not consistent*.

As shown by Vapnik and Chervonenkis [255], consistency holds if and only if convergence in probability of R_{emp} to R is *uniform*, meaning that as the training set increases the probability that $R_{\text{emp}}(w)$ approximates $R(w)$ uniformly tends to 1 on the whole $\mathscr{W}$. Necessary and sufficient conditions for the consistency of the ERM principle is the finiteness of the *Vapnik-Chervonenkis dimension* (VC-dimension) of the hypothesis space $\mathscr{H}$.

The VC-dimension of the hypothesis space $\mathscr{H}$ is, loosely speaking, the largest number of examples that can be separated into two classes in all possible ways by the set of functions $f(x, w)$. The VC-dimension h measures the complexity and descriptive power of the hypothesis space and is often proportional to the number of free parameters of the model $f(x, w)$.

Vapnik and Chervonenkis also provide bounds on the deviation of the empirical risk from the expected risk. A bound that holds with probability $(1 - p)$ is the following one:

$$R(w) \leq R_{\text{emp}}(w) + \sqrt{\frac{h\left(\ln \frac{2\ell}{h} + 1\right) - \ln \frac{p}{4}}{\ell}}, \quad \forall w \in \mathscr{W}.$$

By analyzing the bound, if one neglects logarithmic factors, in order to obtain a small expected risk, both the empirical risk *and* the ratio h/ℓ between the VC-dimension of the hypothesis space and the number of examples have to be small. In other words, a valid generalization after training is obtained if the hypothesis space is sufficiently powerful to allow reaching a small empirical risk, i.e., to learn correctly the training examples, but not too powerful to simply memorize the training examples without extracting the structure of the problem. For a larger model flexibility, a larger number of examples is required to achieve a similar level of generalization.

The choice of an appropriate value of the VC-dimension h is crucial to get good generalization performance, especially when the number of data points is limited.

The method of *structural risk minimization* (SRM) has been proposed by Vapnik based on the above bound, as an attempt to overcome the problem of choosing an appropriate value of h. For the SRM principle, one starts from a nested structure of hypothesis spaces

$$\mathcal{H}_1 \subset \mathcal{H}_2 \subset \cdots \subset \mathcal{H}_n \subset \cdots \tag{7.13}$$

with the property that the VC-dimension $h(n)$ of the set $\mathcal{H}_n$ is such that $h(n) \leq h(n+1)$. As the subset index n increases, the minima of the empirical risk decrease, but the term responsible for the confidence interval increases. The SRM principle chooses the subset $\mathcal{H}_n$ for which minimizing the empirical risk yields the best bound on the actual risk. Disregarding logarithmic factors, the following problem must be solved:

$$\min_{\mathcal{H}_n} \left(R_{\mathrm{emp}}(w) + \sqrt{\frac{h(n)}{\ell}} \right). \tag{7.14}$$

The SVM algorithm described in the following is based on the SRM principle, by minimizing a bound on the VC-dimension and the number of training errors at the same time.

The mathematical derivation of SVMs is summarized first for the case of a linearly separable problem, also to build some intuition about the technique. The notation follows [203].

7.2.5.1 Linearly Separable Problems

Assume that the examples are linearly separable, meaning that there exist a pair (w,b) such that:
$$w \cdot x + b \geq 1 \quad \forall x \in \mathrm{Class}_1,$$
$$w \cdot x + b \leq -1 \quad \forall x \in \mathrm{Class}_2.$$

The hypothesis space contains the functions:

$$f_{w,b} = \mathrm{sign}(w \cdot x + b).$$

Because scaling the parameters (w,b) by a constant value does not change the decision surface, the following constraint is used to identify a unique pair:

$$\min_{i=1,\ldots,\ell} |w \cdot x_i + b| = 1.$$

A structure on the hypothesis space can be introduced by limiting the norm of the vector w. It has been demonstrated by Vapnik that if all examples lie in an

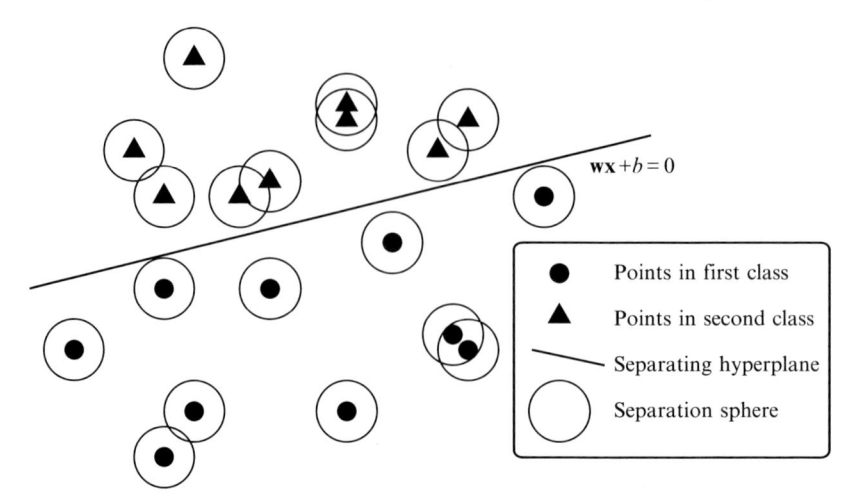

Fig. 7.8 Hypothesis space constraint. The separating hyperplane must maximize the margin. Intuitively, no point has to be too close to the boundary so that some noise in the input data will not ruin the classification

n-dimensional sphere with radius R, then the set of functions $f_{w,b} = \text{sign}(w \cdot x + b)$ with the bound $\|w\| \leq A$ has a VC-dimension h that satisfies

$$h \leq \min\{\lceil R^2 A^2 \rceil, n\} + 1.$$

The geometrical explanation of why bounding the norm of w constrains the hypothesis space is as follows (see Fig. 7.8): If $\|w\| \leq A$, then the distance from the hyperplane (w,b) to the closest data point has to be larger than $1/A$, because one considers only hyperplanes that do not intersect spheres of radius $1/A$ placed around each data point. In the case of linear separability, minimizing $\|w\|$ amounts to determining a separating hyperplane with the maximum *margin* (distance between the convex hulls of the two training classes measured along a line perpendicular to the hyperplane).

The problem can be formulated as:

$$\text{Minimize}_{w,b} \; \tfrac{1}{2}\|w\|^2,$$
$$\text{Subject to} \quad y_i(w \cdot x_i + b) \geq 1 \; i = 1, \ldots, \ell.$$

The problem can be solved by using standard quadratic programming (QP) optimization tools.

The dual quadratic program, after introducing a vector $\Lambda = (\lambda_1, \ldots, \lambda_\ell)$ of nonnegative Lagrange multipliers corresponding to the constraints is as follows:

$$\text{Maximize}_\Lambda \; \Lambda \cdot 1 - \tfrac{1}{2} \Lambda \cdot D \cdot \Lambda,$$
$$\text{Subject to} \quad \begin{cases} \Lambda \cdot y = 0, \\ \Lambda \geq 0 \end{cases} \tag{7.15}$$

7.2 Techniques

where y is the vector containing the example classification, and D is a symmetric $\ell \times \ell$ matrix with elements $D_{ij} = y_i y_j x_i \cdot x_j$.

The vectors x_i for which $\lambda_i > 0$ are called *support vectors*. In other words, support vectors are the ones for which the constraints in (7.15) are active. If w^* is the optimal value of w, the value of b at the optimal solution can be computed as $b^* = y_i - w^* \cdot x_i$ for any support vector x_i, and the classification function can be written as

$$f(x) = \text{sign}\left(\sum_{i=1}^{\ell} y_i \lambda_i^* x \cdot x_i + b^*\right).$$

Note that the summation index can as well be restricted to support vectors, because all other vectors have null λ_i^* coefficients. The classification is determined by a linear combination of the classifications obtained on the examples y_i weighted according to the scalar product between input pattern and example pattern (a measure of the "similarity" between the current pattern and example x_i) and by parameter λ_i^*.

7.2.5.2 Nonseparable Problems

If the hypothesis set is unchanged but the examples are not linearly separable, one can introduce a penalty proportional to the constraint violation ξ_i (collected in vector Ξ), and solve the following problem:

$$\text{Minimize}_{w,b,\Xi} \; \frac{1}{2}\|w\|^2 + C\left(\sum_{i=1}^{\ell} \xi_i\right)^k,$$

$$\text{Subject to} \begin{cases} y_i(w \cdot x_i + b) \geq 1 - \xi_i, & i = 1,\dots,\ell \\ \xi_i \geq 0, & i = 1,\dots,\ell \\ \|w\|^2 \leq c_r, \end{cases} \quad (7.16)$$

where the parameters C and k determine the cost caused by constraint violation, while c_r limits the norm of the coefficient vector. In fact, the first term to be minimized is related to the VC-dimension, while the second is related to the empirical risk (see the above-described SRM principle). In our case, k is set to 1.

7.2.5.3 Nonlinear Hypotheses

Extending the above techniques to nonlinear classifiers is based on mapping the input data x into a higher-dimensional vector of *features* $\varphi(x)$ and using *linear* classification in the transformed space, called the *feature space*. The SVM classifier becomes:

$$f(x) = \text{sign}\left(\sum_{i=1}^{\ell} y_i \lambda_i^* \varphi(x) \cdot \varphi(x_i) + b^*\right).$$

After introducing the *kernel function* $K(x,y) \equiv \varphi(x) \cdot \varphi(y)$, the SVM classifier becomes

$$f(x) = \text{sign}\left(\sum_{i=1}^{\ell} y_i \lambda_i^* K(x,x_i) + b^*\right),$$

and the quadratic optimization problem becomes

$$\text{Maximize}_\Lambda \ \Lambda \cdot 1 - \tfrac{1}{2}\Lambda \cdot D \cdot \Lambda,$$

$$\text{Subject to} \quad \begin{cases} \Lambda \cdot y = 0 \\ 0 \leq \Lambda \leq C1, \end{cases}$$

where D is a symmetric $\ell \times \ell$ matrix with elements $D_{ij} = y_i y_j K(x_i, x_j)$.

An extension of the SVM method is obtained by weighting in a different way the errors in one class with respect to the error in the other class, for example, when the number of samples in the two classes is not equal, or when an error for a pattern of a class is more expensive than an error on the other class. This can be obtained by setting two different penalties for the two classes: C^+ and C^- so that the function to minimize becomes

$$\frac{1}{2}\|w\|^2 + C^+ \left(\sum_{i:y_i=+1}^{\ell} \xi_i\right)^k + C^- \left(\sum_{i:y_i=-1}^{\ell} \xi_i\right)^k.$$

If the feature functions $\varphi(x)$ are chosen with care, one can calculate the scalar products without actually computing all features, therefore greatly reducing the computational complexity.

For example, in a one-dimensional space one can consider monomials in the variable x multiplied by appropriate coefficients a_n:

$$\varphi(x) = (a_0 1, a_1 x, a_2 x^2, \ldots, a_d x^d),$$

so that $\varphi(x) \cdot \varphi(y) = (1+xy)^d$. In more dimensions, it can be shown that if the features are monomials of degree $\leq d$, then one can always determine coefficients a_n so that:

$$K(x,y) = (1 + x \cdot y)^d.$$

The kernel function $K(\cdot,\cdot)$ is a convolution of the canonical inner product in the feature space. Common kernels for use in an SVM are the following.

1. Dot product: $K(x,y) = x \cdot y$; in this case no mapping is performed, and only the optimal separating hyperplane is calculated.
2. Polynomial functions: $K(x,y) = (x \cdot y + 1)^d$, where the *degree d* is given.
3. Radial basis functions (RBF): $K(x,y) = e^{-\gamma \|x-y\|^2}$, with parameter γ.
4. Sigmoid (or neural) kernel: $K(x,y) = \tan h(ax \cdot y + b)$, with parameters a and b.
5. ANOVA kernel: $K(x,y) = \left(\sum_{i=1}^{n} e^{-\gamma(x_i-y_i)}\right)^d$, with parameters γ and d.

When ℓ becomes large, the quadratic optimization problem requires an $\ell \times \ell$ matrix for its formulation, and so it rapidly becomes an unpractical approach as the training set size grows. A decomposition method where the optimization problem is split into an active and an inactive set is introduced in [203]. The work in [149] introduces efficient methods to select the working set and to reduce the problem by taking advantage of the small number of support vectors with respect to the total number of training points.

7.2.5.4 Support Vectors for Regression

Support vector methods can be applied also for regression, i.e., to estimate a function $f(x)$ from a set of training data $\{(x_i, y_i)\}$. As it was the case for classification, one starts from the case of linear functions and then preprocesses the input data x_i into an appropriate feature space to make the resulting model nonlinear.

To fix the terminology, the linear case for a function $f(x) = w \cdot x + b$ can be summarized. The convex optimization problem to be solved becomes:

$$\text{Minimize}_w \ \tfrac{1}{2}\|w\|^2,$$
$$\text{Subject to} \ \begin{cases} y_i - (w \cdot x_i + b) \leq \varepsilon \\ (w \cdot x_i + b) - y_i \leq \varepsilon, \end{cases}$$

assuming the existence of a function that approximates all pairs with ε precision.

If the problem is not feasible, a *soft margin* loss function with slack variables ξ_i, ξ_i^*, collected in vector Ξ, is introduced in order to cope with the infeasible constraints, obtaining the following formulation [256]:

$$\text{Minimize}_{w,b,\Xi} \ \frac{1}{2}\|w\|^2 + C\left(\sum_{i=1}^{\ell} \xi_i^* + \sum_{i=1}^{\ell} \xi_i\right),$$

$$\text{Subject to} \ \begin{cases} y_i - w \cdot x_i - b \leq \varepsilon - \xi_i^*, & i = 1, \ldots, \ell \\ w \cdot x_i + b - y_i \leq \varepsilon - \xi_i, & i = 1, \ldots, \ell \\ \xi_i^* \geq 0, & i = 1, \ldots, \ell \\ \xi_i \geq 0, & i = 1, \ldots, \ell \\ \|w\|^2 \leq c_r. \end{cases} \quad (7.17)$$

As in the classification case, C determines the tradeoff between the flatness of the function and the tolerance for deviations larger than ε. Detailed information about support vector regression can also be found in [238].

7.2.6 Nearest Neighbor's Methods

Although SVMs and MLPs are widely used because of their practical effectiveness, in some cases their implementation can be excessively complex if a small number of training examples is provided. Furthermore, the training time can be large in some cases and adding new examples is problematic because all model parameters are going to be influenced.

A basic form of learning, also known as instance-based learning or case-based or memory-based, works as follows. The examples are stored and no action is taken until a new input pattern demands an output value. When this happens, the memory is searched for examples that are *near* the new input pattern, the vicinity being measured by a suitable metric.

Over a century old, this form of data mining is still being used very intensively by statisticians and machine learners alike. Let us explore nearest weighted k-nearest-neighbor technique(WKNN).

Let $k \leq \ell$ be a fixed positive integer; consider a feature vector x. A simple algorithm to estimate its corresponding outcome y is the following:

1. Find within the training set the k indices $i_1, \ldots, i_k$ whose feature vectors $x_{i_1}, \ldots, x_{i_k}$ are nearest (according to a given feature-space metric) to the given x vector.
2. Calculate the estimated outcome y by the following average, weighted with the inverse of the distance between feature vectors:

$$y = \frac{\sum_{j=1}^{k} \frac{y_{i_j}}{d(x_{i_j}, x) + d_0}}{\sum_{j=1}^{k} \frac{1}{d(x_{i_j}, x) + d_0}}, \qquad (7.18)$$

where $d(x_i, x)$ is the distance between the two vectors in the feature space (for example, the Euclidean distance), and d_0 is a small real constant used to avoid division by zero.

The WKNN algorithm is simple to implement, and it often achieves low estimation errors. Its main drawbacks are the algorithmic complexity of the testing phase and, from a more theoretical point of view, the high VC dimension. A simpler technique, called KNN, does not use distance-dependent weights: The output is just a simple average of the outputs corresponding to the k closest examples. Of course, the quick and dirty version is for $k = 1$; the output is simply that of the closest example in memory.

Kernel methods and locally weighted regression, considered in the previous sections, can be seen as flexible and smooth generalizations of the nearest neighbor's idea; instead of applying a brute exclusion of the distant point, all points contribute to the output but with a significance ("weight") related to their distance from the query point.

7.3 Selecting Features

As in all scientific challenges, the development of models with predicting power has to start from appropriate measurements, statistics, *input features*. The literature

7.3 Selecting Features

for selecting features is very rich in the areas of pattern recognition, neural networks, and machine learning. Before starting to learn a parametric or nonparametric model from the examples, one must be sure that the input data (input features) have sufficient information to predict the outputs. This qualitative criterion can be made precise in a statistical way with the notion of *mutual information* (MI for short).

An output distribution is characterized by an *uncertainty* that can be measured from the probability distribution of the outputs. The theoretically sound way to measure the uncertainty is with the *entropy*; see below for the precise definition. Now, after one knows a specific input value x, the uncertainty in the output can decrease. The amount by which the uncertainty in the output decreases after the input is known is termed *mutual information*.

If the mutual information between a feature and the output is zero, knowledge of the input does not reduce the uncertainty in the output. In other words, the selected feature cannot be used (in isolation) to predict the output – no matter how sophisticated our model is. The MI measure between a vector of input features and the output (the desired prediction) is therefore very relevant to identify promising (informative) features. Its use in feature selection is pioneered in [19].

Let us now come to some definitions for the case of a classification task where the output variable c identifies one among N_c classes and the input variable x has a finite set of possible values, see Fig. 7.9. For example, one may think about predicting whether a run of an algorithm on an instance will converge (class 1) or not (class 0) within the next minute. Among the possible features extracted from the data one would like to obtain a highly-informative set, so that the classification problem starts from sufficient information, and only the actual construction of the classifier is left.

At this point one may ask: Why not use the raw data instead of features? For sure there is no better way to use all possible information. True, but the *curse of dimensionality* holds here: If the dimension of the input is too large, the learning task becomes unmanageable. Think for example about the difficulty of estimating probability distributions from samples in very high dimensional spaces. Heuristically, one aims at a small subset of features, possibly close to the smallest possible, which contains sufficient information to predict the output.

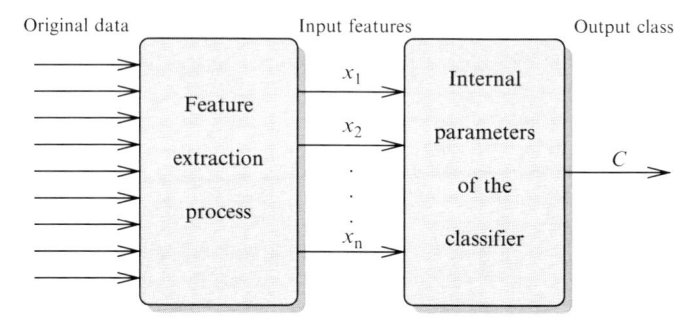

Fig. 7.9 A classifier mapping input features extracted from the data to an output class

7.3.1 Correlation Coefficient

Let Y be the random variable associated with the output classification and let $\Pr(y)$ ($y \in Y$) be the probability of y being its outcome; X_i is the random variable associated to the input variable x_i, and X is the input vector random variable, whose values are x.

The most widely used measure of *linear relationship* is the Pearson product–moment *correlation coefficient*, which is obtained by dividing the covariance of the two variables by the product of their standard deviations. In the above notation, the correlation coefficient $\rho_{X_i,Y}$ between the ith input feature X_i and the classifier's outcome Y, with expected values μ_{X_i} and μ_Y and standard deviations σ_{X_i} and σ_Y, is defined as:

$$\rho_{X_i,Y} = \frac{\text{cov}[X_i, Y]}{\sigma_{X_i} \sigma_Y} = \frac{E[(X_i - \mu_{X_i})(Y - \mu_Y)]}{\sigma_{X_i} \sigma_Y}, \quad (7.19)$$

where E is the expected value of the variable and cov is the covariance. After simple transformations one obtains the equivalent formula:

$$\rho_{X_i,Y} = \frac{E[X_i Y] - E[X_i] E[Y]}{\sqrt{E[X_i^2] - E^2[X_i]} \sqrt{E[Y^2] - E^2[Y]}}. \quad (7.20)$$

The correlation value varies from -1 to 1. Correlation close to 1 means increasing linear relationship (an increase of the feature value x_i relative to the mean is usually accompanied by an increase of the outcome y), close to -1 means a decreasing linear relationship. The closer the coefficient is to zero, the weaker the correlation between the variables, for example, the plot of (x_i, y) points looks like an isotropic cloud around the expected values, without an evident direction.

As mentioned before, statistically independent variables have zero correlation, but zero correlation does not imply that the variables are independent. The correlation coefficient detects only *linear* dependencies between two variables: It may well be that one variable has full information and actually determines the value of the second, as in the case that $y = f(x_i)$, while still having zero correlation.

7.3.2 Correlation Ratio

In many cases, the desired outcome of our learning algorithm is categorical (a "yes/no" answer or a limited set of choices). The correlation coefficient assumes that the outcome is quantitative, thus it is not applicable to the categorical case. To sort out general dependencies, the *correlation ratio* method [158] can be used.

The basic idea behind the correlation ratio is to partition the sample feature vectors into classes according to the observed outcome. If a feature is significant, then it should be possible to identify at least one outcome class where the feature's average value is significantly different from the average on all classes, otherwise that component would not be useful to discriminate any outcome.

Suppose that one has a set of ℓ sample feature vectors, possibly collected during previous stages of the algorithm that one is trying to measure. Let ℓ_y be the number of times that outcome $y \in Y$ appears, so that one can partition the sample feature vectors by their outcome:

$$\forall y \in Y \quad S_y = \left((x_{jy}^{(1)}, \ldots, x_{jy}^{(n)}); j = 1, \ldots, \ell_y\right).$$

In other words, the element $x_{jy}^{(i)}$ is the ith component (feature) of the jth sample vector among the ℓ_y samples having outcome y. Let us concentrate on the ith feature from all sample vectors, and compute its average within each outcome class:

$$\forall y \in Y \quad \bar{x}_y^{(i)} = \frac{\sum_{j=1}^{\ell_y} x_{jy}^{(i)}}{\ell_y}$$

and the overall average:

$$\bar{x}^{(i)} = \frac{\sum_{y \in Y} \sum_{j=1}^{\ell_y} x_{jy}^{(i)}}{\ell} = \frac{\sum_{y \in Y} \ell_y \bar{x}_y^{(i)}}{\ell}.$$

Finally, the *correlation ratio* between the ith component of the feature vector and the outcome is given by

$$\eta_{X_i,Y}^2 = \frac{\sum_{y \in Y} \ell_y (\bar{x}_y^{(i)} - \bar{x}^{(i)})^2}{\sum_{y \in Y} \sum_{j=1}^{\ell_y} (x_{jy}^{(i)} - \bar{x}^{(i)})^2}.$$

If the relationship between values of the i-th feature component and values of the outcome is linear, then both the correlation coefficient and the correlation ratio are equal to the slope of the dependence:

$$\eta_{X_i,Y}^2 = \rho_{X_i,C}^2.$$

In all other cases, the correlation ratio can capture nonlinear dependencies.

7.3.3 Entropy and Mutual Information

As already mentioned, the statistical uncertainty in the output class is measured by its *entropy*:

$$H(Y) = -\sum_{y \in Y} \Pr(y) \log \Pr(y). \tag{7.21}$$

Let us now evaluate the impact that the ith input feature x_i has on the classifier's outcome y. The entropy of Y after knowing the input feature value ($X_i = x_i$) is:

$$H(Y|x_i) = -\sum_{y \in Y} \Pr(y|x_i) \log \Pr(y|x_i),$$

where $\Pr(y|x_i)$ is the conditional probability of being in class y given that the ith feature has value x_i.

Finally, the *conditional entropy* of the variable Y is the expected value of $H(Y|x_i)$ over all values $x_i \in X_i$ that the ith feature can have:

$$H(Y|X_i) = E_{x_i \in X_i}\left[H(Y|x_i)\right] = -\sum_{x_i \in X_i} \Pr(x_i) H(Y|x_i). \tag{7.22}$$

The conditional entropy $H(Y|X_i)$ is always less than or equal to the entropy $H(Y)$. It is equal if and only if the ith input feature and the output class are statistically independent, i.e., the joint probability $\Pr(y, x_i)$ is equal to $\Pr(y)\Pr(x_i)$ for every $y \in Y$ and $x_i \in X_i$. The amount by which the uncertainty decreases is by definition the *mutual information* $I(X_i; Y)$ between variables X_i and Y:

$$I(X_i; Y) = I(Y; X_i) = H(Y) - H(Y|X_i). \tag{7.23}$$

An equivalent expression that makes the symmetry between X_i and Y evident is:

$$I(X_i; Y) = \sum_{y, x_i} \Pr(x_i, y) \log \frac{\Pr(y, x_i)}{\Pr(y)\Pr(x_i)}. \tag{7.24}$$

Although very powerful theoretically, estimating mutual information for a high-dimensional feature vector starting from labeled samples is not a trivial task. A heuristic method that uses only mutual information between individual features and the output is presented in [19], using Fraser's algorithm [102].

Let us note that the mutual information measure is different from correlation measures. A feature can be informative even if not linearly correlated with the output, and the mutual information measure does not even require the two variables to be quantitative. A categorical variable (sometimes called a nominal variable) is one that has two or more categories, but there is no intrinsic ordering to the categories. For example, gender is a categorical variable with two categories (male and female) and no intrinsic ordering.

To make it short: Trust the correlation coefficient only if you have reasons to suspect *linear* relationships, otherwise other correlation measures are possible, in particular the correlation ratio can be computed even if the outcome is not quantitative. Finally, use mutual information to estimate arbitrary dependencies between qualitative or quantitative features!

7.4 Applications

Let us now present some application of supervised learning applied to optimization so that the above-described techniques can be observed in action.

Prediction is used to improve combinatorial optimization search in [50, 52]. In particular, given a local search algorithm A, one learns to predict the final outcome of A, the f value determined at convergence, when started from a given initial configuration X. In other words, one wants to estimate the value of a point when used as *initial point* for a local search. Suitable features are extracted from the configuration, and some sample trajectories are generated so that an initial database relating features of the starting point and final results (local minima identified by A) is then fed to a machine learning technique for building a regression model (linear or nonlinear). If algorithm A is Markov, i.e., if the future of A's search trajectory depends only on the current state and not on the past history, not only the starting states but also all other states along each search trajectory may be used as training data. After a trained model $\text{VALUE}_A(X)$ is available, the STAGE algorithm [50, 52] alternates between hill-climbing phases guided by the original function to optimize f, and hill-climbing phases guided by the new function $\text{VALUE}_A(X)$. The explanation is that, after an initial solution is exploited by local search A, a new and hopefully better initial point is searched for in its vicinity. The method can be used while optimizing a single instance, but also for different instances with similar statistical structure. In this case a single model is trained on an instance and then *transferred* to other similar instances. Sample applications and cases of *transfer* are studied in the cited paper. The function $\text{VALUE}_A(X)$ has another interpretation as *value function* in the Reinforcement Learning method; see also Chap. 8.

Locally-weighted regression for memory-based stochastic optimization is considered in [188] to accelerate the convergence of stochastic optimization. In stochastic optimization, each input X does not produce a single output $f(X)$ but a *distribution* of output values; this is the standard case when the output is obtained through a physical process and measured through real-world noisy instruments. The context is that of optimizing industrial plants in which each experiment is very costly. The experiments are therefore mined to build a global nonlinear model of the expected output, including confidence bars. The method is related to the historic *response surface methodology* and *kriging*, see also Chap. 6 about model-based search.

In other applications of supervised learning, one aims at predicting the running time distribution of a heuristic or stochastic optimization algorithm. Running times of heuristic algorithms are not known a priori; in fact, some instances are empirically harder than others to solve. The reason is that heuristics try to exploit some peculiar characteristic of the problem structure that is not necessarily present at the same degree in every instance. Some heuristics that are effective on some instances can significantly worsen their convergence times on other instances. Exact algorithms also can have a huge variance in their running times. For example, let us consider a branch and bound algorithm that solves exactly instances of an NP-complete problem. If the instance is an YES instance, then the algorithm might find a solution before visiting the exponentially high number of possible states, but if the instance is a NO instance, then all states have to be considered and the running time can increase markedly. Stochastic branch and bound algorithms perform random choices to break ties, and randomized algorithms may shuffle the instance input data before solving. In these cases the choice on the random seed can significantly impact the running time.

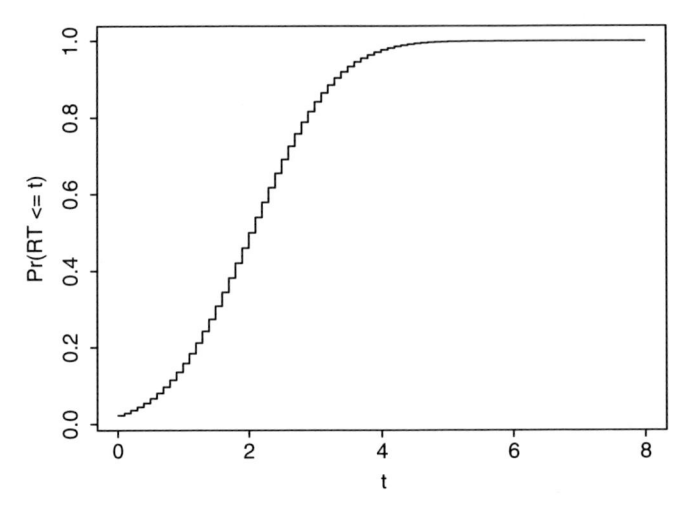

Fig. 7.10 Runtime distribution of an algorithm. The termination probability increases with the running time t. In the case depicted above, the probability for the algorithm to terminate before time t is $\Pr(RT \leq t) = \int_0^t \frac{1}{\sigma\sqrt{2\pi}} \exp-\frac{(x-\mu)^2}{2\sigma^2} dx$ with $\mu = 2$ and $\sigma = 1$

Let RT_{A,s_i} be the running time for an algorithm A to terminate successfully on an input s_i. The *run time distribution* (RTD) is the cumulative probability distribution of the continuous random variable RT_{A,s_i}.

$$RTD_A(t) = \Pr(RT_{A,s_i} \leq t)$$

An example of runtime distribution can be seen in Fig. 7.10.

Supervised machine learning techniques can be used to learn a model $\widehat{RT_A}(\phi(s_i)), \mathbf{w})$ for the runtime of an algorithm A, depending on features $\phi(s_i)$ extracted from a problem instance and on suitable parameters $\mathbf{w}$.

In [171] the authors use linear and nonlinear regression models to predict runtime distributions of an exact branch-and-bound algorithm for solving an instance of a quadratic assignment problem (QAP). The experimental approach suggested for a single algorithm is as follows:

1. Select a problem instance distribution.
2. Select a set of inexpensive features.
3. Generate a set of example instances. For each one, run the algorithm to compute the running time and compute the required features.
4. Use machine learning to determine a function of the features to predict the running time.
5. Fine-tune the feature selection by eliminating redundant features.

For the specific application of the QAP considered, the running time for instances of the same size can vary by several orders of magnitude. The aim of the chapter is to use linear and nonlinear regression techniques to predict the run time and to determine which instance features are more relevant for the prediction. Once the problem size is fixed, the authors select 35 features, which are the input of the regression. The logarithm of the running time is the output. The selected features range from statistical measures on the problem instance to measures on graphs representing relation between features. The importance of a feature for the determination of the hardness of an instance is determined by measuring the cost of its omission in terms of the root-mean-squared error (RMSE). Once the run-time distribution for different algorithms is learned, it is possible to predict their running time on new instances and select the algorithm that performs better on a per-instance basis.

In [199] models of the empirical hardness of random SAT instances are developed for three different SAT solvers and for two different distributions of instances: uniform random 3-SAT with varying ratio of clauses-to-variables, and uniform random 3-SAT with fixed ratio of clauses-to variables. Surprisingly accurate models are built in all cases. Furthermore, the models are analyzed to determine which features are most useful in predicting whether an instance will be hard to solve. The input parameters of the model are 91 selected instance features. Among the features considered, there is the problem size, the ratio between the number of clauses and variables, several statistical measures on traces of short local search probes, and measures derived from graph representations of SAT instances. Both linear and nonlinear (quadratic) models are learned by considering the most promising combinations of the instance features. The models are used to produce very difficult instances to solve by rejection sampling, and to build SATzilla, an algorithm portfolio for SAT that selects the best algorithm for an instance using the empirical hardness models described earlier.

A naïve Bayes classifier is a simple probabilistic classifier based on applying Bayes' theorem with strong (naive) independence assumptions: The probability for having a set of features, given a class C, is obtained by the product of the individual probabilities $Pr(feature_i|C)$, probabilities that are estimated from the relative frequencies of occurrence in the example data. The classification is then executed by choosing the class with the largest estimated posterior probability of the class given the measured features.

In [60] the authors use a "low-knowledge" Bayesian classifier, which does not rely on complex prediction models, to build a portfolio that combines three different algorithms for solving a job scheduling optimization problem. A model is learned that predicts the quality of the solution after t seconds and classifies an algorithm in the *best* or *behind* classes if it performs better than the others or not. The input parameters used are just three instance features selected from the problem domain and the output is the quality of a solution after a time t and how far it is from the best-known solution for the instance. They also propose a racing algorithm that allocates computational resources to each algorithm in a manner proportional to their quality, by using Bayesian learners to predicting which algorithm will perform better within the remaining running time.

In case of parametric algorithms, run times can vary noticeably with the change of single parameters. Therefore, another application of run-time distribution prediction is automatic parameter tuning. By using a regression technique, one can for each parameter configuration c find the run-time distribution of algorithm $A[c]$, and fit a model. Once the models are trained, they have to be evaluated on the new instance to find the best configuration c^*, which minimizes the predicted running time. The problem here is the curse of dimensionality. Learning a linear model is not expensive, but the number of models to learn explodes with the number of parameters and their realizations. If algorithm A has L parameters and each of them has M realizations, or in case of continuous parameters it is discretized in M values, then the number of configurations is $C = M^L$, a number that rapidly explodes. The solution proposed in [143] consists of including the algorithm parameters and their combinations in the inputs of the models. There is only a single model that is learned for every parameter configuration. Since the parameters are input of the model, the discretization of continuous parameters is avoided, and it is possible to use various optimization techniques to find the parameter set that minimizes the learned RTD.

Different models such as *Bayesian networks* are considered in [138] to predict the run time and control search and reasoning algorithms. Bayesian networks, also known as *belief networks*, are graphical models that represent a set of variables and their probabilistic interdependencies. For example, probabilistic relationships between diseases and symptoms can be represented. Given symptoms, the network computes the probabilities of various diseases. In the cited application, the learned models predict the ultimate length of a trial by observing the behavior of the search algorithm during an early phase. The models are then used for dynamic run-time decisions, for example, to decide about restarts. Additional discussions about *reactive* restarts are presented in Sect. 9.5. Automated PAC learning models for discovering optimization algorithms are reviewed in [54]: Algorithm tuning is executed by *learning* algorithms, which automatically learn domain-specific procedures from example runs. The review contains examples in the area of backtracking and dynamic programming.

Interesting summaries of statistical machine learning approaches for large-scale optimization are present in [13]. Up to now we considered applications of supervised machine learning techniques for optimization. It is of interest to note that one can travel also in the opposite direction: Issues of configuration and parameter tuning are also present for machine learning algorithms. The problem of finding parameter settings that will result in optimal performance of a given learning algorithm is studied for example in [163]: Their *wrapper* technique explores the space of parameter values by running the basic algorithm many times on training and holdout sets produced by cross-validation. The final parameter setting are therefore tuned for the specific induction algorithm and dataset being studied.

7.4.1 Learning a Model of the Solver

In the model-based framework introduced by Chap. 6, let us see how a supervised machine learning technique can be used to enhance a local search algorithm's awareness of the overall landscape shape. The searcher is modeled as a function mapping the *starting point* of the search to the *best point* found in the ensuing trajectory.

7.4 Applications

The following building blocks will be introduced in this section: a local search heuristic, a way to describe it as a stochastic function, a statistical tool to evaluate this model, and the combination of these elements.

7.4.1.1 The Reactive Affine Shaker

Let us begin with the single, most important building block: The searcher. It must be intelligent, and it must run as fast as possible toward a minimum.

The Reactive Affine Shaker heuristic [57], based on the Affine Shaker proposed in [32], is a self-tuning local search algorithm where no prior knowledge is assumed on the function to be minimized and only evaluations at arbitrary values of the independent variables are allowed. The Reactive Affine Shaker maintains a small "search region" $\mathscr{R}$ around the current point x.

The Reactive Affine Shaker is able to reshape the search region $\mathscr{R}$ according to the occurrence or lack of success during the last step: If a step in a certain direction yields a better objective value, then $\mathscr{R}$ is expanded along that direction; it is reduced otherwise. Therefore, once a promising direction is found, the probability that subsequent steps will follow the same direction is increased, and the search proceeds more and more aggressively in that direction until bad results reduce its prevalence. The algorithm is outlined in Fig. 7.11.

The algorithm starts with an isotropic search region centered around the initial point (line 2). Next, new trail points are repeatedly generated (line 4). If the resulting point $x + \Delta$ yields a lower objective value (line 5 and following), then the current

---

f Function to minimize
x Initial point
$\mathscr{R}$ Search region
Δ Current displacement

---

1. **function** Reactive_Affine_Shaker (f, x)
2. $\mathscr{R} \leftarrow$ small isotropic set around x
3.   **while** (local termination condition is not met)
4.   Pick $\Delta \in \mathbb{R}^d$ such that $x + \Delta, x - \Delta \in \mathscr{R}$
5.   **if** $f(x + \Delta) < f(x)$
6. $x \leftarrow x + \Delta$;
7.   Extend $\mathscr{R}$ along Δ
8.   Center $\mathscr{R}$ on x
9.   **else if** $f(x - \Delta) < f(x)$
10. $x \leftarrow x - \Delta$;
11.   Extend $\mathscr{R}$ along Δ
12.   Center $\mathscr{R}$ on x
13.   **else**
14.   Reduce $\mathscr{R}$ along Δ
15.   **return** x;

Fig. 7.11 The Reactive Affine Shaker algorithm

position is updated and $\mathscr{R}$ is expanded along the direction of Δ. To increase the probability of finding a better point, if $x+\Delta$ does not lead to an improvement, also $x-\Delta$ is checked (line 9 and following). If neither point improves the objective value, then the search region is reduced along the direction of Δ (line 14) and the current position is not updated. This sequence of steps is repeated until a local termination condition is verified. Common criteria to terminate the search are the number of iterations, the size of the search region (if too small, it indicates that no precise direction for improvement can be detected; therefore, the system is already close to a local minimum), and a large number of iterations without further improvement.

To keep an acceptable level of complexity, the search region is implemented as a box defined by d independent vectors $(b_1 \ldots b_d)$, where d is the number of dimensions of the search domain. Shape modifications are implemented as affine transformations of these vectors, as described in the following equation:

$$\forall j \quad b_j \leftarrow \left(I + (\rho - 1)\frac{\Delta \Delta^\mathrm{T}}{\|\Delta\|^2}\right) b_j \qquad (7.25)$$

The affine transformation depends on a dilation parameter ρ, whose value can be heuristically set at 1.2. Notice that the Reactive Affine Shaker has a very small memory footprint: All previous history information is stored in the box vectors.

Figure 7.12 exemplifies the Reactive Affine Shaker dynamics when the function to be minimized is two-dimensional. Two trajectories (ABC and A'B'C') are plotted,

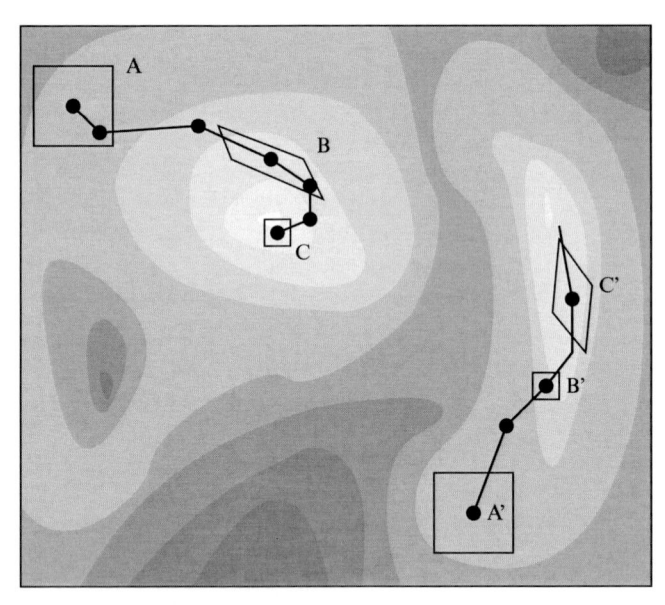

Fig. 7.12 Reactive Affine Shaker geometry: Two search trajectories leading to two different local minima. The evolution of the search regions is also illustrated; starting from points A and A' respectively, the box is enlarged when better points are identified, and elongated along the successful directions, it is compressed when the search is unsuccessful, for example, when close to the local optimum (as in C) or in a difficult place (as in B')

together with the search regions (parallelograms) for some points along the search trajectories. Updates in the search region, operated by means of (7.25), increase the search speed when steps are successful (points A and A' in Fig. 7.12), and reduce it only if no better point is found after the double shot. When a point is close to a local minimum, the repeated reduction of the search frame produces a very fast convergence of the search (point C in Fig. 7.12). Note that another cause of reduction for the search region can be a narrow descent path (a "canyon," such as in point B' of Fig. 7.12), where only a small subset of all possible directions improves the function value. However, once an improvement is found, the search region grows in the promising direction, causing a faster movement along that direction.

7.4.1.2 Modeling the Solver as a Stochastic Function

Let f be a real-valued function defined over a limited domain $D \subset \mathbb{R}^d$. Let L be a local optimization heuristic, and let L_f the algorithm obtained by applying L to function f. L_f works by starting from an initial point $x_1 \in D$ and generating a trajectory $(x_1, \ldots, x_N)$, where N is the number of steps the algorithm performs before a termination condition is verified. If one treats the initial point x_1 as an independent variable (i.e., not randomly generated by the algorithm itself, but fed as a parameter), the algorithm L_f can be seen as a function mapping the initial point of the trajectory to the smallest function value found along the trajectory:

$$L_f : D \to \mathbb{R} \qquad (7.26)$$
$$x_1 \mapsto \min_{i=1,\ldots,N} f(x_i).$$

Note that, since L is a stochastic heuristic relying on random choices, the trajectory is stochastic too, and L_f must be regarded as a stochastic function.

7.4.1.3 An LWR Model of the Stochastic Local Search Transformation

The stochastic function L_f models the transformation executed by local search, from an initial point to the local minimum point in a given attraction basin. After some runs of local search have been executed, one begins to derive knowledge about the structure of the search space, about which region is mapped to which local minima, and about a possible large-scale structure of the local minima showing the way to the most promising areas. Of course, the so-called "no free lunch" theorems of global optimization [269] imply that these techniques will not be effective for general functions (for sure they will not be effective if the value at one point is not related to values at nearby points), but most optimization problems of real interest are indeed characterized by a rich structure that can be profitably mined.

The integration proposed in [58] considers the LWR to model the transformation executed by L_f, therefore to evaluate the potential of future initial points to lead to promising local minima. For each run of the stochastic local search, the

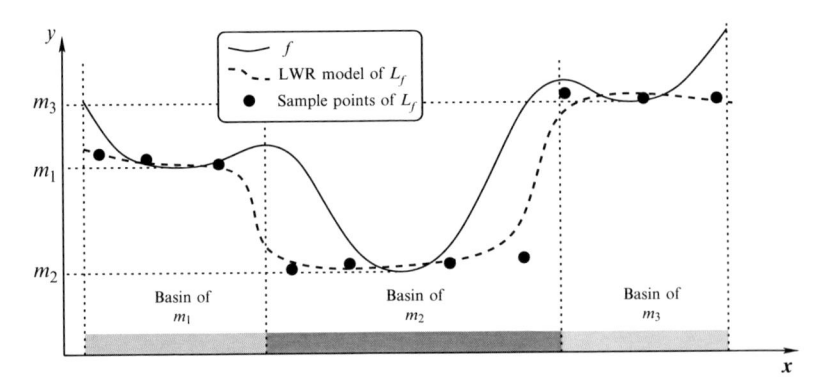

Fig. 7.13 Modeling the local search algorithm L_f

memory-based model will be mined to identify the next initial point. Other options are possible, such as the consideration of an LWR model for describing the original function f. This second hypothesis is considered in [147].

To visualize the effect of the L_f transformation and the related modeling by LWR, Fig. 7.13 describes the application of an LWR technique to L_f in order to model it. Function f has three local minima, whose values are represented as m_1, m_2, and m_3, with m_2 as the global minimum value. Black dots represent sample points, of the form $(x, L_f(x))$, i.e., each is obtained by generating an initial value x, feeding it to the local search algorithm, and retrieving the minimum value of f found along the subsequent trajectory. If the search algorithm makes local moves, as is the case with the Reactive Affine Shaker, the sample points will approximately outline a stepwise function, constant in every attraction basin corresponding to a given local minimum. Let us note that the smooth approximation to the stepwise function is actually useful to give the algorithm a direction to follow to reach promising areas, while an exact constant model on the plateau would not give such direction hint. The LWR model, shown in thick dashed line, is a smoothed-out version of this stepwise function.

The LWR model of L_f (derived in Fig. 7.13) is in turn minimized in order to find the best suitable starting point for the subsequent run of L_f, as shown in Fig. 7.14.

7.4.1.4 The MRAS Heuristic

Figure 7.15 presents the pseudo-code for the memory-based Reactive Affine Shaker (MRAS) heuristic. The parameters are the function f to be minimized and the number of initial sample points in the model. Since we are using the Bayesian version of LWR with prior coefficient distribution, we are not forced to insert into the model a minimum number of points before it becomes useful.

The model g is initially empty; after a sufficient number of sample points are added, it can be evaluated at a query point as a real-valued function. The Reactive Affine Shaker local search algorithm is described in Fig. 7.11.

7.4 Applications

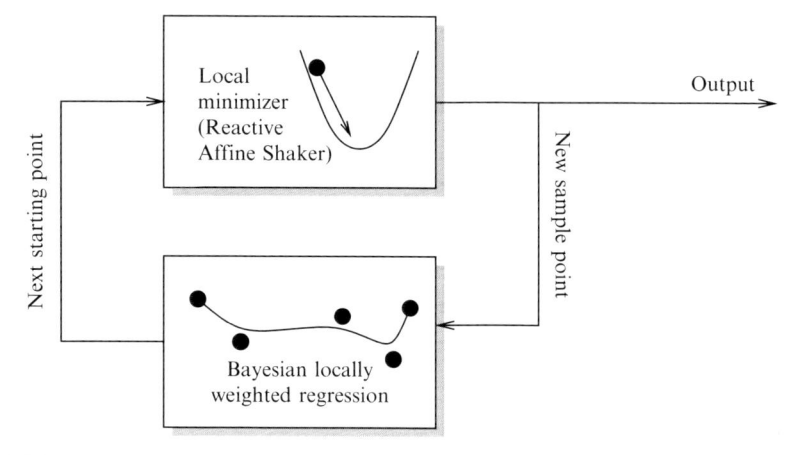

Fig. 7.14 Interaction between Reactive Affine Shaker and the model of L_f: The MRAS algorithm

f	Function to minimize
D	Domain of f
g	B-LWR model of L_f, initially empty
n	Number of initial sample points in g

1. **function** BLWR_Affine_Shaker (f, n)
2. **for** $i \leftarrow 1$ to n
3. $x \leftarrow$ random point in D
4. $x' \leftarrow$ Reactive_Affine_Shaker (f, x)
5. g.addSamplePoint $(x, f(x'))$
6. **while** (termination condition is not met)
7. $x \leftarrow$ Repeated_Affine_Shaker (g)
8. $x' \leftarrow$ Reactive_Affine_Shaker (f, x)
9. g.addSamplePoint $(x, f(x'))$
10. **return** best point found

Fig. 7.15 The MRAS heuristic

Lines 2–5 populate the B-LWR model with a number of sample points, each of the form $(x, L_f(x))$, by repeatedly generating random points in the domain, following a Reactive Affine Shaker trajectory starting from that point (line 4) and storing the result according to the definition (7.26).

Once the model g is populated, the algorithm proceeds by alternating model minimizations and objective function minimizations (lines 6–9). A promising starting point can be found by minimizing g with a multiple-run Reactive Affine Shaker starting from a random point (line 7). The point is used to begin the minimization trajectory for f (line 8). Finally, the result of the optimization run (in terms of initial point, best value in trajectory) is stored into g in order to refine it for the next run.

Chapter 8
Reinforcement Learning

> *That which made me succeed with the bicycle was precisely what had gained me a measure of success in life-it was the hardihood of spirit that led me to begin, the persistence of will that held me to my task, and the patience that was willing to begin again when the last stroke had failed. And so I found high moral uses in the bicycle and can commend it as a teacher without pulpit or creed. He who succeeds, or, to be more exact in handing over my experience, she who succeeds in gaining the mastery of such an animal as Gladys, will gain the mastery of life, and by exactly the same methods and characteristics...*
> *(Frances E. Willard, 1895,* How I Learned to Ride the Bicycle*)*

8.1 Reinforcement Learning Basics: Learning from a Critic

Various opportunities for learning exist both in the real human life and in the theoretical systems defined to study the different contexts. Possibly the most standard form is learning by imitating a *teacher*, this form is called *supervised learning* in the machine learning literature, and has been the subject of Chap. 7.

However, supervised learning is not the only possible royal road to knowledge. In spite of the obsessive "supervising instinct" of parents, kids have an amazing tendency to try out very dangerous situations, like placing fingers into electrical plugs, touching hot lightbulbs, eating horrible things, riding bikes, skates, etc. often out of reach of any teacher. How do they (hopefully) survive? *Reinforcement learning* (RL) is the technical term for a learning context where there is no guiding teacher showing the correct way, but *feedback signals from the environment* that are used by the learner to modify its future actions. Think about bicycle riding, after some initial trials with positive or negative rewards, in the form of admiring friends or injuries to biological tissues, by magic the goal is accomplished. When you are adult, consider a truck and trailer backup, something that cause headaches also to the authors of this book (hopefully because of lack of sufficient learning, and not because of intrinsic inabilities).

The learning problem is now much more difficult: one has to make a *sequence of decisions* (e.g., about the steering wheel rotation). The outcome of each decision is not always fully predictable. In addition to an immediate *reward*, each action causes a change in the system state and therefore a different context for the next decisions. To complicate matters, the reward is often delayed and one aims at maximizing not the immediate reward, but some form of average reward over a sequence of decisions. This means that greedy policies, for example trying to immediately move as close as possible to the parking space, do not always work. In fact, it can be better to go for a smaller immediate reward if this action leads to a state of the systems where bigger rewards can be obtained in the future. *Goal-directed learning from interaction* with an (unknown) environment, with trial-and-error search and delayed reward is the main feature of RL.

Two classes of problems arise in this context: *Prediction* problems, where the policy is fixed and the long-term reward has to be estimated, and the more difficult *control* problems, where the optimal policy maximizing long-term reward has to be discovered. Having an algorithm for a generic *reinforcement learning* problem would be for most computer scientists equivalent to the philosopher's stone for the old alchemists. Just define the system and your goal, or identify a proper simulator, sit down and relax! As you can imagine, things are not so easy.

In this summary we will:

- Define *Markov Decision Processes*, the arena where RL takes place.
- Mention exact solutions when the system model is known, and the number of possible states and possible actions is limited. *Dynamic programming* (DP) is the standard solution engine.
- Focus on the most interesting and practically relevant case where the *curse of dimensionality* (the number of system states times the number of actions is too large to be handled) and the *curse of modeling* (the lack of detailed knowledge about the system behavior) force one to abandon exact techniques in favor of approximations, simulations, production of sample trajectories, and RL strategies.

In Sect. 8.2 we will then consider some relationships between RL and intelligent optimization.

Suggested books to expand the knowledge about RL are [247] and [42] about neuro-dynamic programming, a definition that refers to the relationship with dynamic programming and neural network approximation architectures. For a sample of recent developments, see [169] and [207].

8.1.1 Markov Decision Processes

We want to model *sequential decision making under uncertainty*, with states, decisions, unexpected results, and "long-term" goals to be reached.

A standard Markov process is given by a set of states S with transitions between them described by probabilities $p(i, j)$. The Markov property means that knowledge of the previous states is irrelevant for predicting the probability of subsequent states, transitions depends only on the current state and not on the trajectory executed to reach it, the system is *memory-less*. Markov processes already deal with uncertainty through their use of probabilities, but they lack the notion of decisions, *actions* taken depending on the current state and leading to a different state and to an immediate *reward*. When we add the two ingredients to our arena, we obtain a *Markov Decision Process (MDP)*.

For brevity, we consider only infinite-horizon MDPs with exponentially discounted rewards. An MDP is defined by the tuple (S, A, TRANS, REW, γ), where $S = \{s_1, s_2, \ldots, s_n\}$ is a finite set of states; $A = \{a_1, a_2, \ldots, a_m\}$ is a finite set of

8.1 Reinforcement Learning Basics: Learning from a Critic

actions; $\text{TRANS}(s,a,s')$ is the probability of a transition from state s to s' when action a is taken; $\text{REW}(s,a,s')$ is the reward (a real number) for a transition from s to s' under action a. The *discount factor* γ is used to discount future rewards. It can be interpreted in economic terms: If reward is money, cash flow arriving in the future is not as rewarding as ready cash, and it has to be discounted. The *expected reward* for a given state s and action a is:

$$\text{REW}(s,a) = \sum_{s' \in S} \text{TRANS}(s,a,s') \text{REW}(s,a,s').$$

Now let the intelligent actor come into play. The decision-making component is called a *policy*, and it performs the choice of an action at every step of the process. A (stationary) policy $\pi(a;s)$ is a set of probability distributions that specify, for each state s, the probability that action a is chosen. The policy is deterministic if only one action is selected for each state; in such case, we prefer to write $a = \pi(s)$.

The intelligent component goal is to pick actions at states to maximize a measure of the total reward accumulated during an infinite chain of decisions (infinite horizon). In detail, let us define the *value function* $V^\pi(s)$ of a policy π as the expected, discounted, total reward when following policy π after starting from the initial state s:

$$V^\pi(s) = E\left(\sum_{t=0}^{\infty} \gamma^t r_t \big| s_0 = s\right),$$

where r_t is the reward obtained at step t of the above described process. The intelligent component aims at determining the best policy, the one reaching the optimal value function $V^*(s)$ such that:

$$V^*(s) = \min_\pi V^\pi(s).$$

In addition, let us define the *state-action value function* $Q^\pi(s,a)$ of any policy π (also called *action value function* for brevity) as the expected, discounted, total reward when taking action a in state s and following policy π thereafter:

$$Q^\pi(s,a) = E\left(\sum_{t=0}^{\infty} \gamma^t r_t \big| s_0 = s, a_0 = a\right).$$

The exact $Q^\pi(s,a)$ values for a fixed policy can be obtained, at least for sufficiently small systems, by solving the linear system of Bellman equations:

$$Q^\pi(s,a) = \text{REW}(s,a) + \gamma \sum_{s' \in S} \text{TRANS}(s,a,s') \sum_{a' \in A} \pi(a';s') Q^\pi(s',a').$$

The interpretation is that the value of selecting action a in state s is given by the value of the immediate reward plus the value of the future rewards which one gets by following policy π from the new state. These have to be discounted by γ (they are a step in the future w.r.t. starting immediately from the new state) and properly

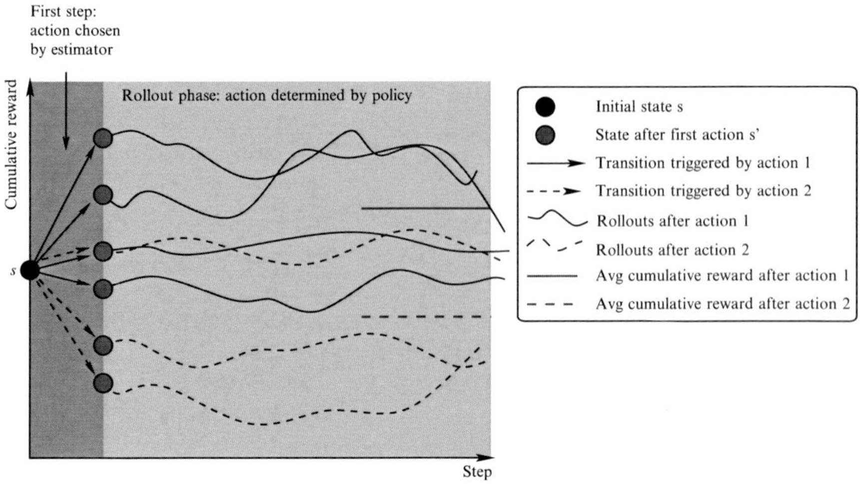

Fig. 8.1 Action value function estimation on state s and two actions, 1 and 2, by multiple rollouts. *Solid lines* refer to the estimation of $Q^\pi(s,1)$, and *dashed lines* show the estimation of $Q^\pi(s,2)$

weighted by transition probabilities and action-selection probabilities given the stochasticity in the process.

Let us note that the vector of values to be determined has dimension equal to the product of states times actions $|S| \times |A|$. Furthermore, a model of the system is required, i.e., full knowledge of $\text{TRANS}(s,a,s')$ and $\text{REW}(s,a)$. When the system is too large, or the model is not completely available, approximations in the form of *reinforcement learning* come to the rescue. As an example, if a *generative model* is available, i.e., a black box that takes state and action in input and produces the reward and next state as output, one can estimate $Q^\pi(s,a)$ through *rollouts*. To estimate $Q^\pi(s,a)$ the generator is used to simulate action a followed by a sufficiently long chain of actions dictated by policy π. The process is repeated several times because of the inherent stochasticity, and averages are calculated. Figure 8.1 provides an example where multiple rollouts are performed to estimate the value function for two possible actions, 1 and 2, with initial state s: solid lines indicate the chains used to estimate $Q^\pi(s,1)$, while dashed lines show the steps to estimate $Q^\pi(s,2)$.

The above described state value function V and state-action value function Q, or approximations thereof, are instrumental in the basic methods of dynamic programming and reinforcement learning.

8.1.2 Dynamic Programming

The central equation for the value function is the Bellman equation:

$$V^*(s) = \max_{a \in A} \left(\text{REW}(s,a) + \gamma \sum_{s' \in S} \text{TRANS}(s,a,s') V^*(s') \right). \quad (8.1)$$

8.1 Reinforcement Learning Basics: Learning from a Critic

The interpretation is intuitive. Given action a, one gets the immediate reward plus the (discounted) value corresponding to the new state, averaged if the transition is not deterministic but stochastic. The best value from a stating state s is obtained by picking the action maximizing the weighted sum.

We skip all mathematical details here and assume that the optimal policy exists and that π is optimal if for all states it reaches the minimum of the right-hand side of Bellman's equation, see [42] for more details.

The basic method to solve Bellman's equation (8.1) is through *value iteration*. One considers one stage of the systems evolution starting from and initial vector of values $V(i), i = 1, ..., n$, calculates the right-hand side of the equation, and substitutes the current value of $V(s)$ with the r.h.s. value. Generally, an infinite number of iterations is required. Let us note that updates of $V(s)$ values can be asynchronous: One does not need to substitute all values simultaneously, provided that all components are considered infinitely often.

A prominent example where the iteration terminates finitely is the shortest path problem, the reader may remember the Bellman-Ford algorithm for computing single-source shortest paths. In graph theory, the shortest path problem asks for a path between two vertices such that the sum of the weights of its edges is minimized. An example is finding the quickest way to get from one location to another on a road map; in this case, the edges represent segments of road and are weighted by the time needed to travel them. In the *single-source* shortest path problem, one has to find shortest paths from a fixed source vertex v to all other vertices in the graph.

The RL states are here the nodes in the graph, the instantaneous reward is the length of an edge, the action is the choice of the next edge to travel, and there is an absorbing state (the destination) when the system is halted. No discount is present ($\gamma = 1$), the total reward is *minimized*, so that min instead of max must be placed in the equations. Actually, instead of "value," the term "cost-to-go" or simply "distance" is more appropriate here. One starts from infinite distance for all nodes apart from the source (where distance from itself is zero), then one *relaxes* edges repeatedly: If deciding to travel along an edge leads to a smaller estimate of the distance, this value is updated and the now better decision to travel along the just discovered edge is saved. Relaxation is in fact value iteration in the dynamic programming terminology.

The shortest-path example helps in understanding how information is transferred in value iteration. As soon as a single forward stage is executed (choice of best action and transition to the new state), the information is transferred *in the opposite direction*: from the new state to the old one. This is called a *backup*. It is the old state that has to be "informed" that bigger future rewards are obtained by selecting a specific action. In the single-source shortest-paths problem, all nodes are ignorant at the beginning, then the nodes directly connected to the source will get the new information that the source is directly reachable and update their values. Then during the next relaxation the nodes connected to these will get the good news that a noninfinite value can be obtained by traveling toward them, and so on and so forth

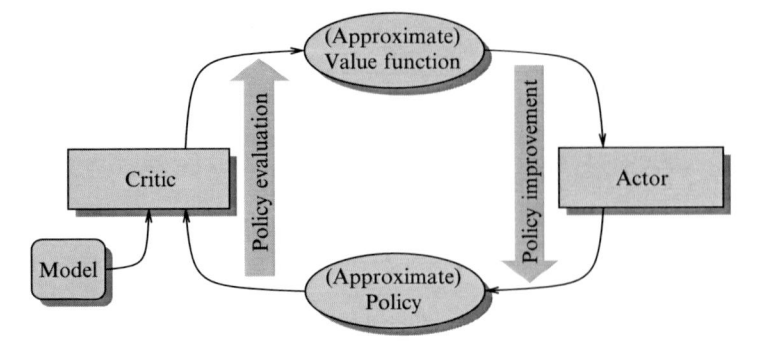

Fig. 8.2 The policy iteration scheme, also known as "Critic-Actor:" a policy is evaluated (critic), the resulting value function is used to define a new policy (actor)

until the nodes farthest from the source will be reached by the good novel about rewards and values!

In general, when an unknown system is tested (either with a simulator or with the "real" evolution), discoveries of large rewards need to be propagated backwards so that previous states are informed and can pick future actions according to the new information.

An alternative to value iteration, which always terminates finitely, is the *policy iteration* method shown in Fig. 8.2, which generates a sequence of policies π_k after starting from an initial proper policy π_0. The first step is *policy evaluation*, which calculates the value function of the current policy π_k through the linear system of equations:

$$V_{\pi_k}(s) = \sum_{s' \in S} \text{TRANS}(s, \pi_k(s), s') \left(\text{REW}(s, \pi_k(s), s') + \gamma V_{\pi_k}(s') \right). \quad (8.2)$$

Let us note that we are evaluating a single policy, so that the max operation is now missing and the resulting equations are indeed linear. The interpretation is the usual one: if $V_{\pi_k}(s)$ is the value function specifying the reward accumulated by starting from a state and following a fixed policy. This has to be equal to the immediate reward obtained plus what can be obtained by starting from the new state, discounted because the "cash" is one step in the future. The second step is *policy improvement*, which computes a new policy by considering the above value function, but by optimizing the first stage, meaning that the arg max operation is back at work, as follows:

$$\pi_{k+1}(s) = \arg\max_{a \in A} \sum_{s' \in S} \text{TRANS}(s, \pi_k(s), s') \left(\text{REW}(s, a, s') + \gamma V_{\pi_k}(s') \right). \quad (8.3)$$

The two steps of evaluation and improvement are repeated until the value function does not change after iterating. If the linear system for policy evaluation is too large to be solved exactly, a limited number of value iteration steps can be used to solve the system approximately between policy improvement steps.

8.1.3 Approximations: Reinforcement Learning and Neuro-Dynamic Programming

The above discussion assumed that the system model (transition probabilities and reward function) is known. In many cases this detailed information is not available but we have a *simulator* of the system, a black box that given current stare and action determines the next state and reward. Alternatively, the real system can be available for experiments. In both cases, more conveniently with a simulator, several sample trajectories can be generated, so that more and more information about the system behavior can be extracted aiming at an optimal control.

A brute force approach can be that of estimating the system model functions, the TRANS(s,a,s') and REW(s,a,s') functions, by executing a very large series of simulations, and then applying the dynamic programming approach previously described.

A second and smarter possibility is to bypass the system model, aiming at learning directly the value functions instrumental to selecting optimal actions. This is the arena of RL. In particular, simulations can be used for policy evaluation, including the *temporal difference* methods such as TD and TD(λ). Finally, the *Q-learning* method can be used to learn an optimal action value function and a derived optimal policy when the system model is not available. To simplify this presentation, we assume that the number of states allows a *lookup table representation* of the system, meaning that all $V(s)$ values can be kept in memory for all states. The problem is not the curse of dimensionality but the curse of modeling.

A possible approach to learn $V_\pi(s)$ is the Monte Carlo one: one generates many sample trajectories for all initial states, accumulates rewards, and estimates averages. A different possibility, called *temporal difference (TD) learning*, acts in an *on-line* and fully incremental manner, by updating estimates of the value function as soon as information about a *single transition* becomes available. Because the TD method bases its update in part on an existing estimate, the method is *bootstrapping*, like Dynamic Programming. The policy π under evaluation runs and, as soon as a new transition from s_t to s_{t+1} is executed, one updates the current value at s_t as follows:

$$V(s_t) \leftarrow V(s_t) + \alpha \left[\underbrace{r_t + \gamma V(s_{t+1})} - V(s_t) \right]. \tag{8.4}$$

Let us motivate: $V(s_t)$ is the old estimate, $r_t + \gamma V(s_{t+1})$ is the new estimate, which is superior because it contains also the fresh information of the just obtained actual reward. Let us note that rewards are primary data in RL, value functions are derived and auxiliary quantities. If the *difference* is zero, there is no motivation to update the old estimate, otherwise the estimate is updated by adding a fraction $\alpha \in [0,1]$ of the difference. It would be unfair to immediately jump to the new value (this is the case if $\alpha = 1$) because the information was obtained from a single sample, without considering transition probabilities to the various states. This is according to the

Robbins-Monro stochastic approximation algorithm to estimate a value from noisy samples. For convergence, α must be decreased in time in an appropriate manner.

When long trajectories are generated with a given policy, a given *difference between prediction and obtained value*, let us call it δ_t, along the trajectory has something to say not only to the state immediately preceding the transition, but also to previous states along the trajectory. For example, if a large δ_t is observed, the value of $V(s_t)$ will increase with TD, causing later increases of $s_{(t-1)}$, which was last updated with the old value, and so on. Aiming at accelerating convergence, one can propagate the information about a given δ_t to the chain of states, which lead to the given transition in the simulation. Because the influence is less and less direct the more we consider states in the past, it is appropriate to discount the difference with a λ value in $[0,1]$. The immediately preceding state s_t value will be updated by summing $\alpha\delta$, the state $s_{(t-1)}$ will be updated by summing $\alpha\lambda\delta$, the state $s_{(t-2)}$ by summing $\alpha\lambda^2\delta$, and so on. This version that aims at bridging the gap between pure TD, a.k.a. TD(0), and full Monte Carlo evaluation is called TD(λ).

TD or TD(λ) can be used for policy evaluation, to be followed be policy improvement, but, if one is impatient, a TD prediction method can be used also for controlling the system: While a policy runs, it is gradually evaluated and improved as soon as updated values become available. One algorithm following this scheme is SARSA, the term derives from the initial letters of State, Action, Reward, State, Action: the policy is chosen according to an action value function $Q(s,a)$. The best action is the one maximizing $Q(s,a)$, in some cases allowing for the *exploration* of seemingly suboptimal actions, let us remember that Q is only an estimate, and two transitions are simulated. Then one updates from the 5-tuple along the trajectory $(s_t, a_t, r_t, s_{t+1}, a_{t+1})$ as:

$$Q(s_t, a_t) \leftarrow Q(s_t, a_t) + \alpha \left[\underbrace{r_t + \gamma Q(s_{t+1}, a_{t+1})} - Q(s_t, a_t) \right]. \quad (8.5)$$

Although in the above example the policy used for generating samples is the same as the policy that is being evaluated, the two can be different in *Q-learning*, which executes an off-policy TD control aiming at approximating the optimal action value function Q^*. In the simplest form the update is:

$$Q(s_t, a_t) \leftarrow Q(s_t, a_t) + \alpha \left[\underbrace{r_t + \gamma \max_{a \in A} Q(s_{t+1}, a)} - Q(s_t, a_t) \right]. \quad (8.6)$$

When the number of states and actions do not permit to store the values in memory, one will have to resort to more compact *approximation architectures* for the value functions and action value functions. For example, a neural network with a compact set of parameters $\mathbf{w}$ can be used to calculate the state value $V(s; \mathbf{w})$, and action values $V(s, a; \mathbf{w})$, see [42].

8.2 Relationships Between Reinforcement Learning and Optimization

Many are the intersections between optimization and DP and RL. First, approximated versions of DP and RL tasks contain in their belly challenging optimization tasks, let us mention the maximization operations in determining the best action when an action value function is available (when the problem is nontrivial because of the huge number of possible actions) or the optimal choice of approximation architectures and parameters in neuro-dynamic programming, not to mention the optimal choice of algorithm details and parameters for a specific RL instance. A recent paper about the interplay of optimization and machine learning (ML) research is [40], which mainly shows how recent advances in optimization can be profitably used in ML.

But the interest of this book goes in the opposite direction: which techniques of RL can be used to improve heuristic algorithms for standard optimization? By standard optimization we mean minimizing a $f(x)$ function, if the optimization task is already a sequential decision problem, no effort is needed. For example, many games, maintenance decisions with limited resources, sequential allocation of communication channels in cellular systems are already in the form appropriate for using DP / RL. Some RL approaches for optimization are also discussed in [13].

An application of RL in the area of local search for solving $\max_x f(x)$ is presented in [51,52]. The rewards from a local search method π starting from an initial configuration x are given by improvements of the best-so-far value f_{best}. In detail, the value function $V^\pi(x)$ of configuration x is given by the expected best value of f seen on a trajectory starting from state x and following the local search method π. In other words, *the value function evaluates the promise of a configuration as a starting state* for π. The curse of dimensionality discourages using directly x, informative features extracted from x are used to compress the state description to a shorter vector $F(x)$, so that the value function becomes $V^\pi(F(x))$. The proposed algorithm STAGE is a smart version of iterated local search, which alternates between two phases. In one, new training data for $V^\pi(F(x))$ are obtained by running local search from the given starting configuration. Let us note that, if local search is memoryless, estimates of $V^\pi(F(x'))$ all points s' along the local search trajectory can be obtained from a single run. In the other phase, one hill-climbs to optimize the value function $V^\pi(F(x))$, instead of the original f, so that a hopefully new and promising starting point is obtained. A suitable approximation architecture $V^\pi(F(x);w)$ is needed, and a supervised ML method to train $V^\pi(F(x);w)$ from the examples. An open issue is the proper definition of features by the researcher, which are chosen by insight in the cited paper. Preliminary results are presented about *transfer*, i.e., the possibility to use a value function learnt on one task for a different task with similar characteristics.

A second application to local search is to supplement f with a "scoring function" to help in determining the appropriate search option at every step. For example, different basic moves or entire different neighborhoods can be applied. RL can in principle make more systematic some of the heuristic approaches involved in designing appropriate "objective functions" to guide the search process. An example is the RL approach to job-shop scheduling in [274, 275], where a neural-network-based $TD(\lambda)$ scheduler is demonstrated to outperform a standard iterative repair (local search) algorithm. The RL problem is designed to pick the best among two possible scheduling actions (moves transforming the current solution). The reward scheme is intended to prefer actions that *quickly* find a *good* schedule. To this end, a fixed negative reinforcement is given for each action leading to a schedule still containing constraint violations. In addition, a scale-independent measure of the final admissible schedule length gives feedback about the schedule quality. Features are extracted from the state, and the δ values along the trajectory are used to gradually update a parametric model $V(s;w)$ of the state value function of the optimal policy, by executing a small step along the gradient direction with length proportional to the "error" value δ. Features are either hand-crafted or determined in an automated manner [275].

Also tree-search techniques can profit from ML. It is well known that variable and value ordering heuristics (choosing the right order of variables or values) can noticeably improve the efficiency of complete search techniques, e.g., for constraint satisfaction problems. For example, RLSAT [168] is a DPLL solver for the satisfiability (SAT) problem, which uses experience from previous executions to learn how to select appropriate branching heuristics from a library of predefined possibilities, with the goal of minimizing the total size of the search tree, and therefore the CPU time. The work [167] extends *algorithm selection for recursive computation*, which is formulated as a sequential decision problem: choosing an algorithm at a given stage requires an immediate cost – in the form of CPU time – and leads to a different state with a new decision to be executed. For SAT, features are extracted from each subinstance to be solved, and TD learning with *least-squares* is used to learn an appropriate action value function, approximated as a linear function of the extracted features. Let us note that some modifications of the basic methods are needed. First, now two new subproblems are obtained instead of one. Second, an appropriate reweighting of the samples is needed to avoid that the huge number of samples close to the leaves of the search tree practically hides the importance of samples close to the root. According to the authors, their work demonstrates that "some degree of reasoning, learning, and decision making on top of traditional search algorithms can improve performance beyond that possible with a fixed set of hand-built branching rules."

A different application is suggested in [42] for the context of constructive algorithms, which build a complete solution by selecting value for a component at a time. One starts from the task data d and repeatedly picks a new index m_n to fix a value u_{m_n}, until values for all N components are fixed. The decision to be made from a partial solution $x_p = (d, m_1, u_{m_1}, ..., m_n, u_{m_n})$ is which index to consider next and which value to assign. Let us assume that K fixed construction algorithms are available for the problem. The application consists of *combining in the most appropriate manner the information obtained by the set of construction algorithms* to fix the next index and value. The approximation architecture suggested is:

8.2 Relationships Between Reinforcement Learning and Optimization

$$V(d,x_p) = \psi_0(x_p, w0) + \sum_{k=1}^{K} \psi_k(x_p, w_k) H_{k,d}(x_p),$$

where $H_{k,d}(x_p)$ is the value obtained by completing the partial solution with the kth method, and ψ are tunable coefficients depending on parameters w, which have to be learned from many runs on different instances of the problem. One evaluates starting points for local search in the previously mentioned STAGE algorithm, while here one evaluates the promise of partial solutions to lead to good complete solutions. Let us note that, in addition to the separate training phase, the K construction algorithms must be run for each partial solution (for each variable-fixing step) to permit to pick the next step in an optimal manner, and therefore this particular application is very hungry for CPU time. A different parametric combination of costs of solutions obtained by two fixed heuristics is considered for the *stochastic programming problem* of maintenance and repair [42]. In the problem, one has to decide whether to immediately repair breakdowns by consuming the limited amount of spare parts, or to keep spare parts for breakdowns at a later phase.

In the context of continuous function optimization, the work in [193] uses RL for replacing a priori defined adaptation rules for the step size in Evolution Strategies with a reactive scheme that automatically adapts step sizes during the optimization process. The states are characterized only by the success rate after a fixed number of mutations, the three possible actions consist of increasing by a fixed multiplicative amount, decreasing, or keeping the current step size. SARSA learning with various reward functions is considered, including combinations of the difference between the current function value and the one evaluated at the last reward computation, and the movement in parameter space (the distance traveled in the last phase). On-the-fly parameter tuning, or *on-line calibration of parameters for evolutionary algorithms* by RL is suggested in [91]. Some of the considered parameters are crossover, mutation, selection operators, and population size. The EA process is divided into episodes, the state describes the main properties of the current population (like mean fitness – or f values – standard deviation, etc.), the reward reflects the progress between two episodes (improvement of the best fitness value), the action consists of determining the vector of control parameters for the next episode.

The trade-off between exploration and exploitation is another issue shared by RL and stochastic local search techniques. In RL an optimal policy is learnt by generating samples, in some cases samples are generated by a policy that is being evaluated and then improved. If the initial policy is generating states only in a limited portion of the entire state space, there will be no way to learn the true optimal policy. A paradigmatic example is the n-armed bandit problem, so named by analogy with a slot machine. In the problem, one is repeatedly faced with a choice among different options, or actions. After each choice, a numerical reward is generated from a stationary probability distribution that depends on the selected action. The objective is to maximize the expected total reward over some time period. Let us note that the action value function in this case depends only on the action, not on the state, which is unique in this simplified problem. The *n-armed bandit problem* is considered in Chap. 10 about racing. In the more general case of many states, no theoretically sound ways exist to combine exploration and exploitation in an optimal manner. One resorts to heuristics, for example to picking actions not in a greedy manner considering an action value function, but based on a simulated-annealing-like probability distribution. Alternatively, actions that are within ε of the optimal value can be chosen with a nonzero probability.

Chapter 9
Algorithm Portfolios and Restart Strategies

Union gives strength.
(Aesop)

9.1 Introduction: Portfolios and Restarts

Let us consider *Las Vegas* algorithms, which always terminate with a correct solution and have a stochastic distribution of the runtime, the time required to terminate. We are interested both in the expected value of the runtime and in its standard deviation. The standard deviation is related to the risk; in some cases having a larger average CPU time with a small deviation is preferable to having a smaller average but with some instances requiring enormous times. There are two simple ways to combine the execution of different algorithms or of different versions of the same algorithm (with different random seeds) to obtain different expected runtimes and standard deviations: One is based on restarting an algorithm if it does not terminate within a given time limit, and the other one is based on combining more runs in a time-sharing interleaving manner: the portfolio approach.

The *algorithm portfolio* method, first proposed in [142], follows the standard practice in economics to obtain different return-risk profiles in the stock market by combining stocks characterized by individual return-risk values. Risk is related to the standard deviation of return. Using an algorithm, portfolio consists of running more algorithms concurrently on a sequential computer, in a time-sharing manner, by allocating a fraction of the total CPU cycles to each of them. The first algorithm to finish determines the termination time of the portfolio, while the other algorithms are stopped immediately after one reports the solution (Fig. 9.1).

It is intuitive that the CPU time can be radically reduced in this manner. To clarify ideas consider an extreme example where, depending on the initial random seed, the runtime can be of 1 s or of 1,000 s, with the same probability. If one runs a single process the expected runtime is approximately of 500 s. If one runs more copies, the probability that at least one of them is lucky (i.e., that it terminates in 1 s) increases very rapidly towards one. Even if termination is now longer than 1 s because more copies share the same CPU, it is intuitive that the expected time will be much shorter than 500.

A portfolio can consist of different algorithms but also of different runs of the same algorithm, with different random seeds. In the case of more runs of the same

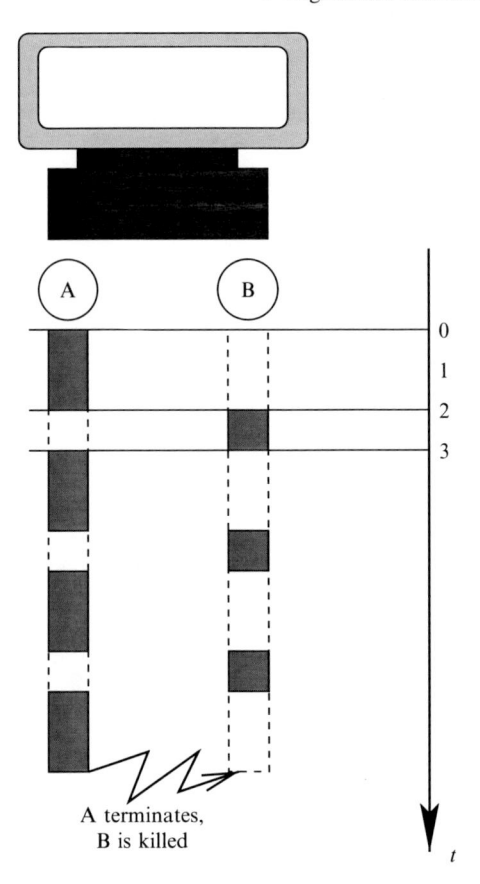

Fig. 9.1 A sequential portfolio strategy

algorithm, there is a different way to have more runs share a given CPU, by terminating a run prematurely and *restarting* the algorithm.

In the above example, a run can be stopped if it does not terminate within 1 s. Because the probability to have a sequence of unlucky cases rapidly goes to zero, again the expected runtime of the restart strategy will be much less than 500 s.

As another example, when surfing the Web the response time to deliver a page can vary a lot. The customary behavior of clicking again on the same link after the patience is finished can save the user from an "endless" waiting time.

9.2 Predicting the Performance of a Portfolio from its Component Algorithms

To make the above intuitive arguments precise let $T_\mathscr{A}$ be the random variable describing the time of arrival of process $\mathscr{A}$ when the whole CPU time is allocated to it. Let $p_\mathscr{A}(t)$ be its probability distribution. The *survival function* $S_\mathscr{A}(t)$ is the

9.2 Predicting the Performance of a Portfolio from its Component Algorithms

probability that process $\mathscr{A}$ takes longer that t to complete:

$$S_{\mathscr{A}}(t) = \Pr(T_{\mathscr{A}} > t) = \int_{\tau > t} p_{\mathscr{A}}(\tau) d\tau = 1 - F_{\mathscr{A}}(t),$$

where $F_{\mathscr{A}}(t)$ is the corresponding cumulative distribution function. If only a fraction α of the total CPU time is dedicated to it in a time-sharing fashion, with arbitrarily small time quanta and no process swapping overhead, we can model the new system as a process $\mathscr{A}'$ whose time of completion is described by random variable $T_{\mathscr{A}'} = \alpha^{-1} T_{\mathscr{A}}$. Its probability distribution and cumulative distribution function are:

$$p_{\mathscr{A}'}(t) = p_{\mathscr{A}}(\alpha t), \quad F_{\mathscr{A}'}(t) = F_{\mathscr{A}}(\alpha t), \quad S_{\mathscr{A}'}(t) = S_{\mathscr{A}}(\alpha t).$$

Consider a portfolio of two algorithms $\mathscr{A}_1$ and $\mathscr{A}_2$. To simplify the notation, let T_1 and T_2 be the random variables associated with their termination times (each being executed on the whole CPU), with survival functions S_1 and S_2. Let α_1 be the fraction of CPU time allocated to process running algorithm $\mathscr{A}_1$. Then the fraction dedicated to $\mathscr{A}_2$ is $\alpha_2 = 1 - \alpha_1$. The completion time of the two-process portfolio system is, therefore, described by the random variable

$$T = \min\{\alpha_1^{-1} T_1, \alpha_2^{-1} T_2\}. \tag{9.1}$$

The survival function of the portfolio is

$$\begin{aligned}
S(t) &= \Pr(T > t) = \Pr(\min\{\alpha_1^{-1} T_1, \alpha_2^{-1} T_2\} > t) \\
&= \Pr(\alpha_1^{-1} T_1 > t \wedge \alpha_2^{-1} T_2 > t) = \Pr(\alpha_1^{-1} T_1 > t) \Pr(\alpha_2^{-1} T_2 > t) \\
&= \Pr(T_1 > \alpha_1 t) \Pr(T_2 > \alpha_2 t) \\
&= S_1(\alpha_1 t) S_2(\alpha_2 t).
\end{aligned}$$

The probability distribution of T can be obtained by differentiation:

$$p(t) = -\frac{\partial S(t)}{\partial t}.$$

Finally, the expected termination value $E(T)$ and the standard deviation $\sqrt{\text{Var}(T)}$ can be calculated.

By turning the α_1 knob, therefore, a series of possible combinations of expected completion time $E(T)$ and risk $\sqrt{\text{Var}(T)}$ becomes available. Figure 9.2 illustrates an interesting case where two algorithms $\mathscr{A}$ and $\mathscr{B}$ are given. Algorithm $\mathscr{A}$ has a fairly low average completion time, but it suffers from a large standard deviation, because the distribution is bimodal or heavy-tailed, while algorithm $\mathscr{B}$ has a higher expected runtime, but with the advantage of a lower risk of having a longer computation. By combining them as described earlier, we obtain a parametric distribution whose expected value and standard deviation are plotted against each other for α_1 going from 0 (only $\mathscr{B}$ executed) to 1 (pure $\mathscr{A}$). Some of the obtained distributions are *dominated* (there are parameter values that yield distributions with lower mean

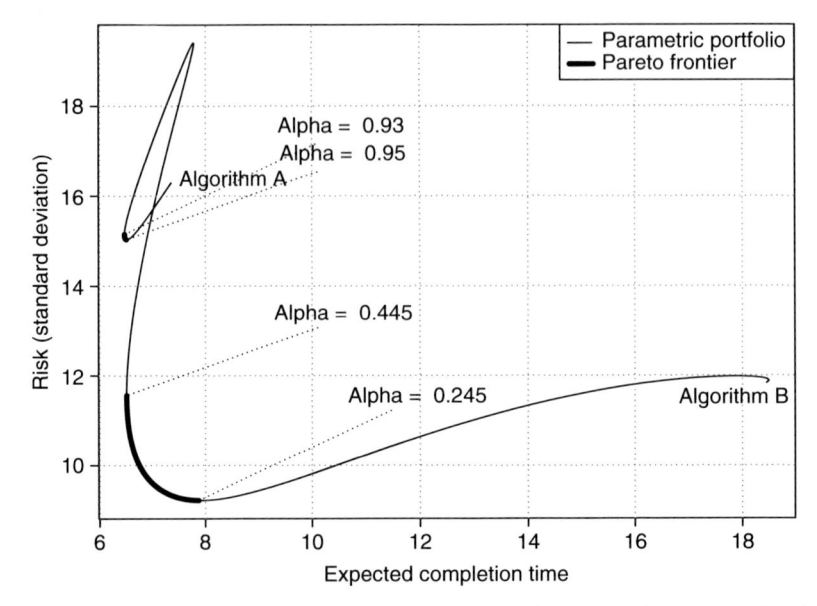

Fig. 9.2 Expected runtime vs. standard deviation (risk) plot. The efficient frontier contains the set of nondominated configurations (a.k.a. Pareto-optimal or extremal points)

time *and* lower risk) and can be eliminated from consideration in favor of better alternatives, while the choice among the nondominated possibilities (on the *efficient frontier* shown in black dots in the figure) has to be specified depending on the user preferences between lower expected time or lower risk. The choice along the Pareto frontier is similar when investing in the stock market: While some choices can be immediately discarded, there are no free meals and a higher return comes with a higher risk.

9.2.1 Parallel Processing

Let us consider a different context [113] and assume that N equal processors and two algorithms are available so that one has to decide how many copies n_i to run of the different algorithms, as illustrated in Fig. 9.3. Of course no processor should remain idle, therefore $n_1 + n_2 = N$.

Consider time as a discrete variable (clock ticks or fractions of second), let T_i be the discrete random variable associated with the termination time of algorithm i having probability $p_i(t)$, the probability that process i halts precisely at time t. As in the previous case, we can define the corresponding cumulative probability and survival functions:

$$F_i(t) = \Pr(T \leq t) = \sum_{\tau=0}^{t} p_i(\tau), \quad S_i(t) = \Pr(T > t) = \sum_{\tau=t+1}^{\infty} p_i(\tau).$$

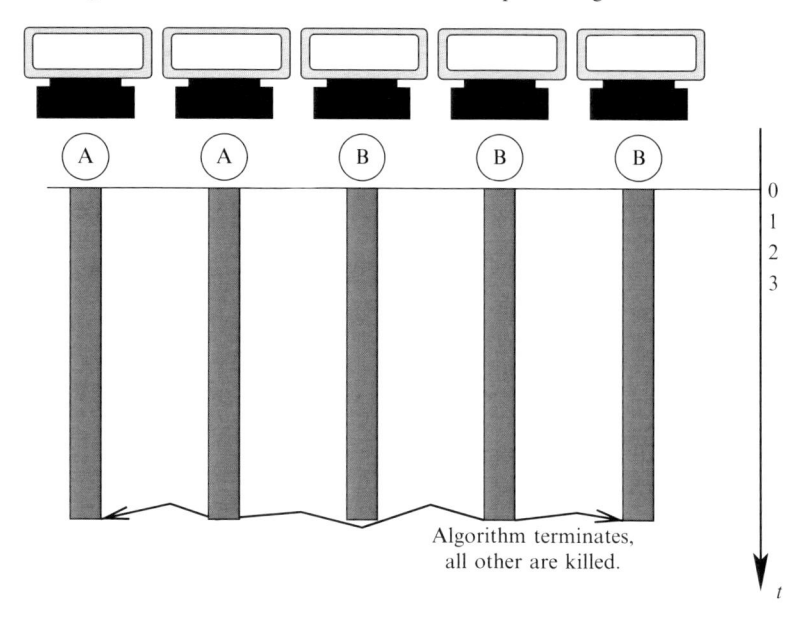

9.2 Predicting the Performance of a Portfolio from its Component Algorithms

Fig. 9.3 A portfolio strategy on a parallel machine

To calculate the probability $p(t)$ that the portfolio terminates exactly at time $T = t$, we must sum probabilities for different events: the event that one processor terminates at t, while the other ones take more than t, the event that two processors terminate at t, while the other ones take more than t, and so on. The different runs are independent and therefore probabilities are multiplied. If $n_1 = N$ (all processors are assigned to the same algorithm), this leads to:

$$p(t) = \sum_{i=1}^{N} \binom{N}{i} p_1(t)^i S_1(t)^{N-i}. \tag{9.2}$$

The portfolio survival function $S(t)$ is easier to compute on the basis of the survival function of the single process $S_1(t)$:

$$S(t) = S_1(t)^N. \tag{9.3}$$

When two algorithms are considered, the probability computation has to be modified to consider the different ways to distribute i successes at time t among the two sets of copies such that $i_1 + i_2 = i$ (i_1 and i_2 being nonnegative integers).

$$p(t) = \sum_{\substack{0 \le i_1 \le n_1 \\ 0 \le i_2 \le n_2 \\ i_1 + i_2 \ge 1}} \binom{n_1}{i_1} p_1(t)^{i_1} S_1(t)^{n_1 - i_1} \binom{n_2}{i_2} p_2(t)^{i_2} S_2(t)^{n_2 - i_2}. \tag{9.4}$$

Similar although more complicated formulas hold for more algorithms. As before, the complete knowledge about $p(t)$ can then be used to calculate the mean and variance of the runtimes. Portfolios can be effective to cure the typical heavy-tailed behavior of $p_i(t)$ in many complete search methods, where very long runs occur more frequently than one may expect, in some cases leading to infinite mean or infinite variance [112]. Heavy-tailed distributions are characterized by a power-law decay, also called tails of Pareto-Lévy form, namely:

$$P(X > x) \approx Cx^{-\alpha},$$

where $0 < \alpha < 2$ and C is a constant.

Experiments with the portfolio approach [113, 142] show that, in some cases, a slight mixing of strategies can be beneficial provided that one component has a relatively high probability of finding a solution fairly quickly. Portfolios are also particularly effective when negatively correlated strategies are combined: one algorithm tends to be good on the cases that are more difficult for the other one, and vice versa. In branch-and-bound applications [113], one finds that ensembles of risky strategies can outperform the more conservative best-bound strategies. In a suitable portfolio, a depth-first strategy that often quickly reaches a solution can be preferable to a breadth-first strategy with lower expected time but longer time to obtain a first solution.

Portfolios can also be applied to component routines inside a single algorithm, for example to determine an acceptable move in a local-search based strategy.

9.3 Reactive Portfolios

The assumption in the above analysis is that the statistical properties of the individual algorithms are known beforehand, so that the expected time and risk of the portfolio can be calculated, the efficient frontier determined and the final choice executed depending on the risk-aversion nature of the user. The strategy is, therefore, off-line: a preliminary exhaustive study of the components precedes the portfolio definition.

If the distributions $p_i(t)$ are unknown, or if they are only partially known, one has to resort to reactive strategies, where the strategy is dynamically changed in an online manner when more information is obtained about the task(s) being solved and the algorithm status. For example, one may derive a maximum-likelihood estimate of $p_i(t)$, use it to define the first values α_i of the CPU time allocations, and then refine the estimate of $p_i(t)$ when more information is received and use it to define subsequent allocations. A preliminary suggestion of dynamic online strategies is present in [142].

Dynamic strategies for search control mechanisms in a portfolio of algorithms are considered in [66, 67]. In this framework, statistical models of the quality of solutions generated by each algorithm are computed online and used as a control

Variable	Scope	Meaning
$\mathscr{A}_i$	(input)	i-th algorithm ($i = 1, \ldots, n$)
b_k	(input)	k-th problem instance ($k = 1, \ldots, m$)
f_P	(input)	Function deciding time slice according to expected completion time
f_τ	(input)	Function estimating the expected completion time based on history
τ_i	(local)	Expected remaining time to completion of current run of algorithm $\mathscr{A}_i$
α_i	(local)	Fraction of CPU time dedicated to algorithm $\mathscr{A}_i$
history	(local)	Collection of data about execution and status of each process

1. **function AOTA**($\mathscr{A}_1, \ldots, \mathscr{A}_n, b_1, \ldots, b_m, f_P, f_\tau$)
2. **repeat** $\forall b_k$
3. ⎡ initialize $(\tau_1, \ldots, \tau_n)$
4. ⎢ **while** (b_k **not** solved)
5. ⎢ ⎡ update $(\alpha_1, \ldots, \alpha_n) \leftarrow f_P(\tau_1, \ldots, \tau_n)$
6. ⎢ ⎢ **repeat** $\forall \mathscr{A}_i$
7. ⎢ ⎢ ⎡ run $\mathscr{A}_i$ for a slot of CPU time $\alpha_i \Delta t$
8. ⎢ ⎢ ⎢ update history of $\mathscr{A}_i$
9. ⎢ ⎢ ⎣ update estimated termination $\tau_i \leftarrow f_\tau$(history)
10. ⎣ update model f_τ considering also the complete history of the last solved instance

Fig. 9.4 The inter-problem AOTA framework

strategy for the algorithm portfolio, to determine how many cycles to allocate to each of the interleaved search strategies.

A "life-long learning" approach for dynamic algorithm portfolios is considered in [104]. The general approach of "dropping the artificial boundary between training and usage, exploiting the mapping *during* training, and including training time in performance evaluation," also termed Adaptive Online Time Allocation [103], is fully in the reactive search spirit. In the inter-problem AOTA framework, see Fig. 9.4, a set of algorithms $\mathscr{A}_i$ is given, together with a sequence of problem instances b_k, and the goal is to minimize the runtime of the whole set of instances. The model used to predict the runtime $p_i(t)$ of algorithm $\mathscr{A}_i$ is updated after each new instance b_k is solved. The portion of CPU time α_i is allocated to each algorithm $\mathscr{A}_i$ in the portfolio through a heuristic function, which is decreasing for longer estimated runtimes τ_i.

9.4 Defining an Optimal Restart Time

Restarting an algorithm at time τ is beneficial if its expected time to convergence is less than the expected additional time to converge, given that it is still running at time τ [189]:

$$E[T] < E[T - \tau | T > \tau]. \tag{9.5}$$

Whether restart is beneficial or not depends of course on the distribution of runtimes. As a trivial example, if the distribution is exponential, restarting the algorithm does not modify the expected runtime.

If the distribution is heavy-tailed, restart easily cures the problem. For example, heavy tails can be encountered if a stochastic local search algorithm such as simulated annealing is trapped in the neighborhood of a local minimizer. Although eventually the optimal solution will be visited, an enormous number of iterations can be spent in the attraction basin around the local minimizer before escaping. Restart is a direct method to escape deep local minima!

If the algorithm is always restarted at time τ, each run corresponds to a Bernoulli trial, which succeeds with probability $P_\tau = \Pr(T \leq \tau)$ – remember that T is the random variable associated with termination time of an unbounded run of the algorithm. The number of runs executed by the restart strategy before success follows a geometric distribution with parameter P_τ, in fact the probability of a success at repetition k is $(1-P_\tau)^{k-1} P_\tau$. The distribution of the termination time T_τ of the restart strategy with restart time τ can be derived by observing that at iteration t one has had $\lfloor t/\tau \rfloor$ restarts and $(t \bmod \tau)$ remaining iterations. Therefore, the survival function of the restart strategy is

$$S_\tau(t) = \Pr(T_\tau > t) = (1-P_\tau)^{\lfloor t/\tau \rfloor} \Pr(T > t \bmod \tau). \tag{9.6}$$

The tail decays now in an exponential manner: the restart portfolio is *not* heavy-tailed.

In general, a restart strategy consists of executing a sequence of runs of a randomized algorithm, to solve a given instance, stopping each run k after a time $\tau(k)$ if no solution is found, and restarting an independent run of the same algorithm, with a different random seed. The optimal restart strategy is uniform, i.e., one in which a constant $\tau_k = \tau$ is used to bound each run [175]. In this case, the expected value of the total runtime T_τ, i.e., the sum of runtimes of the successful run, and all previous unsuccessful runs is equal to:

$$E(T_\tau) = \frac{\tau - \int_0^\tau F(t)\,dt}{F(\tau)}, \tag{9.7}$$

where $F(\tau)$ is the cumulative distribution function of the runtime T for an unbounded run of the algorithm, i.e., the probability that the problem is solved before time τ. The demonstration is simple. For a given cutoff τ, each run succeeds with probability $F(\tau)$ (Bernoulli trials) and the mean number of trials before a successful run is encountered is $1/F(\tau)$. The expected length of each run is:

$$\int_0^\tau t p(t)\,dt + \tau(1-F(\tau)).$$

Consider the cases when termination is within τ or later, so that the run is terminated prematurely. Because $p(t) = F'(t)$, this is equal to:

9.4 Defining an Optimal Restart Time

$$\int_0^\tau tF'(t)dt.$$

The result follows from the fact that:

$$\frac{d}{dt}(tF(t)) = tF'(t) + F(t)$$

and therefore:

$$\int_0^\tau tF'(t)dt + \int_0^\tau F(t)dt = \tau F(\tau)$$

giving (9.7).

In the discrete case:

$$E(T_\tau) = \frac{\tau - \sum_{t<\tau} F(t)}{F(\tau)}. \qquad (9.8)$$

If the distribution is known, an optimal cutoff time can be determined by minimizing (9.7). If the distribution is not known, a universal nonuniform strategy, with cutoff sequence: $(1,1,2,1,1,2,4,1,1,2,4,8,\ldots)$ achieves a performance within a logarithmic factor of the expected runtime of the optimal policy, see [175] for details.

Calculating the runtime distribution can require large amounts of CPU time in case of heavy tails because one has to wait for the termination of very long runs. In this case, the *censored sampling* approach can be used. Censored sampling allows to bound the duration of each experimental run and still exploit the information obtained from the runs that converge before the censoring threshold [197]. Let us model the probability density function as $g(t|\theta)$, θ being the parameter to be identified from the experiments. Without censoring one can determine g by maximizing the likelihood $\mathscr{L}$ of the obtained sequence of termination times $\mathscr{T} = (t_1, t_2, \ldots, t_k)$ given θ:

$$\mathscr{L}(\mathscr{T}|\theta) = \prod_{i=1}^k \mathscr{L}(t_i|\theta) = \prod_{i=1}^k g(t_i|\theta). \qquad (9.9)$$

With censoring, some experimental runs will exceed the cutoff time t_c. In these cases the corresponding multiplicative term in (9.9) is substituted with

$$\mathscr{L}_c(t_c|\theta) = \int_{t_c}^\infty g(\tau|\theta)d\tau = 1 - G(t_c|\theta), \qquad (9.10)$$

where $G(t|\theta)$ is the conditional cumulative distribution function corresponding to g.

One has to decide about a proper cutoff threshold t_c. A way to determine it is to ask for target u on the fraction of terminated runs (uncensored samples), run k experiments in parallel (or with interleaving) and stop as soon as the desired target is reached.

The final receipt is therefore: (1) choose an appropriate parametric model for the runtime distribution, (2) determine the best parameters of the model by maximizing the likelihood, where some terms are substituted with the censored likelihood of (9.10), (3) use the estimated runtime distribution to determine the optimal restart time. Some examples of parametric models are considered in [105].

9.5 Reactive Restarts

Up to now the assumption has been that the only observation that can be used is given by the *length of a run* and that the runs are *independent*. Let us now consider more advanced strategies where at least one of these assumptions is relaxed. Given the results mentioned in the previous section, it looks as if the problem is solved for the complete knowledge case and the zero knowledge case (within a multiplicative constant and logarithmic factor that can be large for practical applications). Actually, the most interesting case is between the two situations, when a partial knowledge is available, which is increasing as soon as more data during a run or during a sequence of runs on related instances become available. Real-time observations about the characteristics of a specific instance and the state of the solver *during a run* permit better results.

In [138, 157] it is shown how to use features capturing the state of a solver during the initial phase of the run to *predict the length of a run*, so that the prediction can be used by dynamic restart policies. Bayesian models to predict the runtime starting from both structural evidence available at the beginning of the run, and execution evidence available during the run (in a reactive manner) are trained in a supervised manner. To be more precise, the discrimination is between long and short runs, i.e., runs longer or shorter than the median. The dynamic policy considered in [138] is as follows:

1. Observe a run for O steps (observation horizon)
2. If the run is not terminated predict whether it will converge in a total of L steps
3. If the prediction is negative, restart immediately, otherwise run up to a total of L steps before restarting.

Because the model is not perfect, an important parameter is the model accuracy A, the probability of a correct prediction. If p_i is the probability of a run ending within i steps, the probability of convergence during a single run is, therefore, $p_O + A(p_L - p_O)$ and the expected number of runs until a solution is found is $E(n) = 1/(p_O + A(p_L - p_O))$. An upper bound on the expected number of steps in a single run can be derived by assuming that runs ending within O steps take exactly O steps, while runs terminating between $O+1$ and $O+L$ steps take exactly L steps. The probability of continuation, taking the limited accuracy into account, is $Ap_L + (1-A)(1-p_L)$. An upper bound on the length of a single run is, therefore, $E_{\text{ub}}(R) = O + (L-O)(Ap_L + (1-A)(1-p_L))$, and an upper bound on the expected time to solve a problem with the above policy is $E(n)E_{\text{ub}}(R)$. The estimate can be now minimized by varying L and the observation horizon. The model is rude; for example, no observations during the steps after O are used, only a bound and not the exact expected number of steps is minimized. In spite of its roughness, significantly superior results of the dynamic

9.5 Reactive Restarts

policy w.r.t. the static one are demonstrated. Three different contexts are defined: In the *single instance* context, one has to solve a specific instance as soon as possible, in the *multiple instance* context, one draws cases from a distribution of instances and has to solve either *any instance* as soon as possible, or *as many instances as possible* for a given amount of time allocated (*max instances* problem), see [138] for details.

The assumption of independence among runs is relaxed in [218]. For example, independence is not valid if more runs are on the same instance picked at the beginning from one of several probability distributions. As an example, consider two distributions, one consisting of instances that are solved in 10 iterations, the other one of instances that are solved in 100 iterations. If an instance is not solved in 10 iterations we know that 100 iterations are needed and restarting would only waste computing cycles. Compare this with the situation of a single distribution with probability 0.5 of converging at iteration 10, probability 0.5 of converging at iteration 1,000, with independence among the runs. Here restarting is clearly useful as shown in Sect. 9.4. The work in [218] considers the context where one among several RTDs is picked at the beginning – without informing the user – and a new sample is extracted at each run from the same distribution (e.g., consider two different distributions corresponding to satisfiable or unsatisfiable instances of SAT). The task is to find the optimal restart policy $(t_1, t_2, \dots)$ but now, after each unsuccessful run, the solver's belief about the source distribution can be updated. The problem of finding the optimal restart policy is formulated as a Markov decision process and solved with dynamic programming, considering both the case in which only the termination time is observed, and the case when other predictors of the distribution can be used, for example the evidence obtained during the run about the fact that a SAT instance is or is not satisfiable.

Chapter 10
Racing

The smart player goes with the winners...
(Popular wisdom)

10.1 Exploration and Exploitation of Candidate Algorithms

Portfolios and restarts are simple ways to combine more algorithms, or more runs of a given randomized strategy, to obtain either a lower expected convergence time or a lower risk (variance), or both.

We have already seen that more advanced reactive strategies can be obtained by using a learning loop *while* the portfolio or restart scheme runs. In this way, some of the portfolio parameters or the restart threshold can take fresh information into account.

A related strategy using a "life-long learning" loop to optimize the allocation of time among a set of alternative algorithms for solving a specific instance is termed *racing*. Running algorithms are like horses: After the competition is started one gets more and more information about the relative performance and periodically updates the bets on the winning horses, which are assigned a growing fraction of the available future computing cycles (Fig. 10.1).

A racing strategy is characterized by two components: (1) the estimate of the future potential given the current state of the search, i.e., given the history of the previous iterations and the corresponding results, (2) the allocation of the future CPU cycles to speedup the overall objective of minimizing a function.

Racing is related to a paradigmatic problem in machine leaning and intelligent heuristics known as the *k-armed bandit problem*. One is faced with a slot machine with k arms, which, when pulled, yields a payoff from a fixed but unknown distribution. One wants to maximize the expected total payoff over a sequence of n trials. If the distribution is known one would immediately pull only the best performing arm. What makes the problems intriguing is that one has to split the effort between *exploration* to learn the different distributions and *exploitation* to pull the best arm, once the winner becomes clear. One is reminded of the critical exploration vs. exploitation dilemma observed in optimization heuristics, but there is an important difference: in optimization one is not interested in maximizing the total payoff but in maximizing *the best pull* (the maximum value obtained by a pull in the sequence). The paper [97] is dedicated to determining a sufficient number of

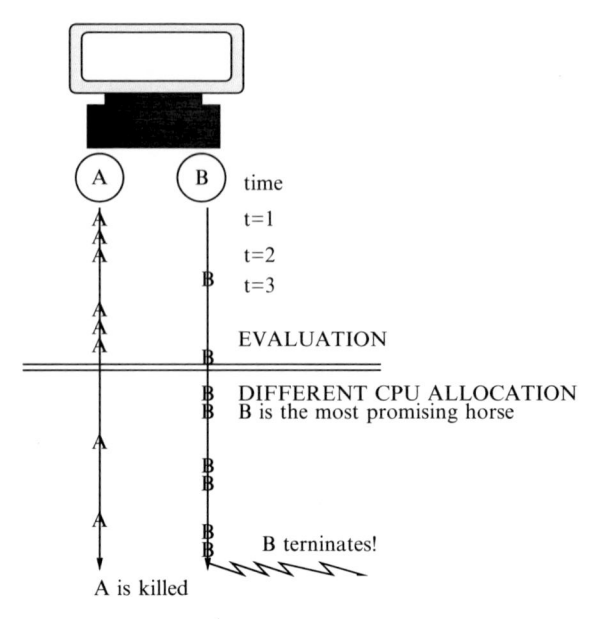

Fig. 10.1 A racing strategy. The different horses (algorithms) are evaluated periodically to reallocate the CPU time shares

pulls to select with a high probability an arm (an hypothesis) whose *average* payoff is near-optimal. The max version of the bandit problem is considered in [67, 68]. An asymptotically optimal algorithm is presented in [243], in the assumption of a generalized extreme value (GEV) payoff distribution for each arm. Our explanation follows closely [242], which presents a simple distribution-free approach.

10.2 Racing to Maximize Cumulative Reward by Interval Estimation

The first algorithm CHERNOFF-INTERVAL-ESTIMATION is for the classical bandit problem, which is then used as a starting point for the THRESHOLD-ASCENT algorithm dedicated to the max k-armed bandit problem. The assumption is that pulling an arm produces a random variable $X_i \in [0, 1]$. Because some effort is spent in exploration to determine (in an approximated manner) the best arm, of course the performance is less than that obtainable by knowing the best arm and pulling it all the time. What one misses by not having the information about the winning horse at the beginning is called *regret*. Precisely, regret is the difference between the payoff obtained by always pulling the best arm on a specific instance minus the cumulative payoff actually received during the racing strategy.

CHERNOFF-INTERVAL-ESTIMATION pulls arms and keeps an estimate of: the number of times n_i of pulls of the ith arm, the expected reward $\bar{\mu}_i = \frac{x_i}{n_i}$ and an

10.2 Racing to Maximize Cumulative Reward by Interval Estimation

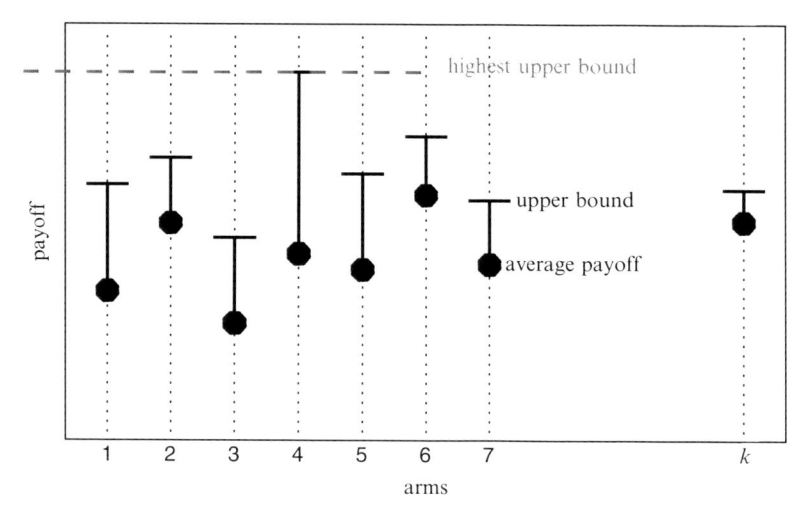

Fig. 10.2 Racing with interval estimation. At each iteration an estimate of the expected payoff of each arm as well as its "error bar" are available

```
1.  function Chernoff_Interval_Estimation(n, δ)
2.  forall i ∈ {1, 2, ..., k} Initialize x_i ← 0, n_i ← 0
3.  repeat n times:
4.      î ← arg max_i U(μ̄_i, n_i)
5.      pull arm î, receive payoff R
6.      x_i ← x_i + R, n_i ← n_i + 1
```

Fig. 10.3 The CHERNOFF-INTERVAL-ESTIMATION routine

upper bound (with a specific minimum probability) on the reward $U(\bar{\mu}_i, n_i)$. At each iteration, the arm with the highest upper bound is pulled (Figs. 10.2 and 10.3). The upper bound is derived from Chernoff's inequality and is as follows:

$$U(\mu, n) = \begin{cases} \mu + \frac{\alpha + \sqrt{2n\mu\alpha + \alpha^2}}{n} & \text{if } n > 0 \\ \infty & \text{otherwise,} \end{cases} \quad (10.1)$$

where $\alpha = \ln\left(\frac{2nk}{\delta}\right)$ and δ regulates our confidence requirements, see later.

Chernoff's inequality estimates how much the empirical average can be different from the real average. Let $X = \sum_{i=1}^{n} X_i$ be the sum of independent identically distributed random variables with $X_i \in [0, 1]$, and $\mu = E[X_i]$ be the real expected value. The probability of an error of the estimate greater than $\beta\mu$ decreases in the following exponential way:

$$P\left[\frac{X}{n} < (1-\beta)\mu\right] < e^{-\frac{n\mu\beta^2}{2}}. \quad (10.2)$$

From this basic inequality, which does not depend on the particular distribution, one derives that, if arms are pulled according to the algorithm in Fig. 10.3, with probability at least $(1-\delta/2)$, for all arms and for all n repetitions the upper bound is not wrong: $U(\bar{\mu}_i, n_i) > \mu_i$. Therefore, each suboptimal arm (with $\mu_i < \mu^*$, μ^* being the best arm expected reward) is not pulled many times and the expected *regret* is limited to at most:

$$(1-\delta)2\sqrt{3\mu^* n(k-1)\alpha} + \delta\mu^* n. \qquad (10.3)$$

A similar algorithm based on Chernoff-Hoeffding's inequality has been presented in a previous work [10]. In their simple UCB1 deterministic policy, after pulling each arm once, one then pulls the arm with the highest bound $U(\bar{\mu}, n_i) = \bar{\mu} + \sqrt{\frac{2\ln n}{n_i}}$, see [10] for more details and experimental results.

10.3 Aiming at the Maximum with Threshold Ascent

Our optimization context is characterized by a set of horses (different stochastic algorithms) aiming at discovering the maximum value for an instance of an optimization problem, for example different greedy procedures characterized by different ordering criteria, see [242] for an application to the Resource Constrained Project Scheduling Problem. The "reward" is the final result obtained by a single run of an algorithm. Racing is a way to allocate more runs to the algorithms that tend to get better results on the given instance.

We are, therefore, not interested in cumulative reward, but in the *maximum* reward obtained at any pull. A way to estimate the potential of different algorithms is to put a threshold *Thres*, and to estimate the probability that each algorithm produces a value above threshold by the corresponding empirical frequency. Unfortunately, the appropriate threshold is not known at the beginning, and one may end up with a trivial threshold – so that all algorithms become indistinguishable – or with an impossible threshold, so that no algorithm will reach it. The heuristic solution presented in [242] reactively learns the appropriate threshold while the racing scheme runs, see Fig. 10.5 and the pseudo-code in Fig. 10.4.

1. **function Threshold_Ascent**(s, n, δ)
2. Thres $\leftarrow 0$
3. **forall** $i \in \{1, 2, ..., k\}$
4.   **forall** R values
5.   Initialize $n_{i,R} \leftarrow 0$
6. **repeat** n times:
7.   **while** (number of payoffs received above threshold $\geq s$)
8. $Thres \leftarrow Thres + \Delta$ (raise threshold)
9. $\hat{i} \leftarrow \arg\max_i U(\bar{v}_i, n_i)$
10.   pull arm $\hat{i}$, receive payoff R
11. $n_{i,R} \leftarrow n_{i,R} + 1$

Fig. 10.4 The THRESHOLD-ASCENT routine

10.4 Racing for Off-Line Configuration of Metaheuristics

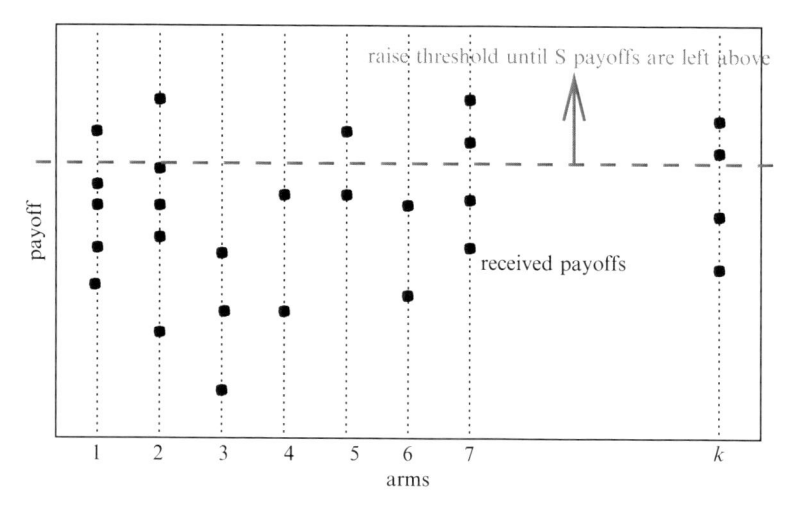

Fig. 10.5 Threshold ascent: the threshold is progressively raised until a selected number of experimented payoffs is left

The threshold starts from zero (remember that all values are bounded in $[0, 1]$), and it is progressively raised until a selected number s of experimented payoffs above threshold is left. For simplicity, but it is easy to generalize, one assumes that payoffs are integer multiples of a suitably small Δ, $R \in \{0, \Delta, 2\Delta, ..., 1-\Delta, 1\}$. In the figure, $\bar{v}_i$ is the frequency with which arm i received a value greater than $Thres$ in the past, an estimate of the probability that it will do so in the future. This quantity is easily calculated from $n_{i,R}$, the number of payoffs equal to R received by horse i. The upper bound U is the same as before.

The parameter s controls the tradeoff between intensification and diversification. If $s = 1$, the threshold becomes so high that no algorithm reaches it: the bound is determined only by n_i and the next algorithm to run is the one with the lowest n_i (Round Robin). For larger values of s, one starts differentiating between the individual performance. A larger s means a more robust evaluation of the different strategies (not based on pure luck – so to speak), but a very large value means that the threshold gets lower and lower so that even poor performers have a chance of being selected. The specific setting of s is, therefore, not so obvious and it looks like more work is needed.

10.4 Racing for Off-Line Configuration of Metaheuristics

The context here is that of selecting in an off-line manner the best configuration of parameters θ for a heuristic solving repetitive problems [44]. Let us assume that the set of possible θ values is finite. For example, a pizza delivery service receives orders and, at regular intervals, has to determine the best route to serve the last customers. In this case, an off-line algorithm tuning (or "configuration"), even if expensive, is worth the effort because it is going to be used for a long time in the future.

There are two sources of randomness in the evaluation: the stochastic occurrence of an instance, with a certain probability distribution, and the intrinsic stochasticity in the randomized algorithm while solving a given instance. Given a criterion $\mathscr{C}(\theta)$ to be optimized with respect to θ, for example the average cost of the route over different instances and different runs, the ideal solution of the configuration problem is:

$$\theta^* = \arg\min_{\theta} \mathscr{C}(\theta), \tag{10.4}$$

where $\mathscr{C}(\theta)$ is the following Lebesgue integral (I is the set of instances, C is the range for the cost of the best solution found in a run, depending on the instance i and the configuration θ):

$$\mathscr{C}(\theta) = E_{I,C}[c(\theta,i)] = \int_I \int_C c(\theta,i) \mathrm{d}P_C(c|\theta,i) \mathrm{d}P_I(i). \tag{10.5}$$

The probability distributions are not known at the beginning. Now, to calculate the expected value for each of the finite configurations, ideally one could use a brute force approach, considering a very large number of instances and runs, tending to infinity. Unfortunately, this approach is tremendously costly, usually each run to calculate $c(\theta,i)$ implies a nontrivial CPU cost, and one has to resort to smarter methods.

First, the above integral in (10.5) is estimated in a Monte Carlo fashion by considering a set of instances. Second, as soon as the first estimates become available, the manifestly poor configurations are discarded so that the *estimation effort is more concentrated onto the most promising candidates*. This process is actually a bread-and-butter issue for researchers in heuristics, with racing one aims at a statistically sound *hands-off* approach. In particular, one needs a statistically sound criterion to determine that a candidate configuration θ_j is *significantly* worse than the current best configuration available, given the current state of the experimentation.

The situation is illustrated in Fig. 10.6, at each iteration a new test instance is generated, and the surviving candidates are run on the instance. The expected performance and error bars are updated. Afterwards, if some candidates have error bars that show a clear inferior performance, they are eliminated from further consideration. Before deciding for elimination, a candidate checks to see whether its optimistic value (top error bar) can beat the pessimistic value of the best performer (Fig. 10.6).

The advantage is clear: costly evaluation cycles to get better estimates of performance are dedicated only to the most promising candidates. Racing is terminated when a single candidate emerges as the winner or when a certain maximum number of evaluations have been executed, or when a target error bar ε has been obtained, depending on available CPU time and application requirements.

The variations of the off-line racing technique depend on the way in which *error bars* are derived from the experimental data.

10.4 Racing for Off-Line Configuration of Metaheuristics

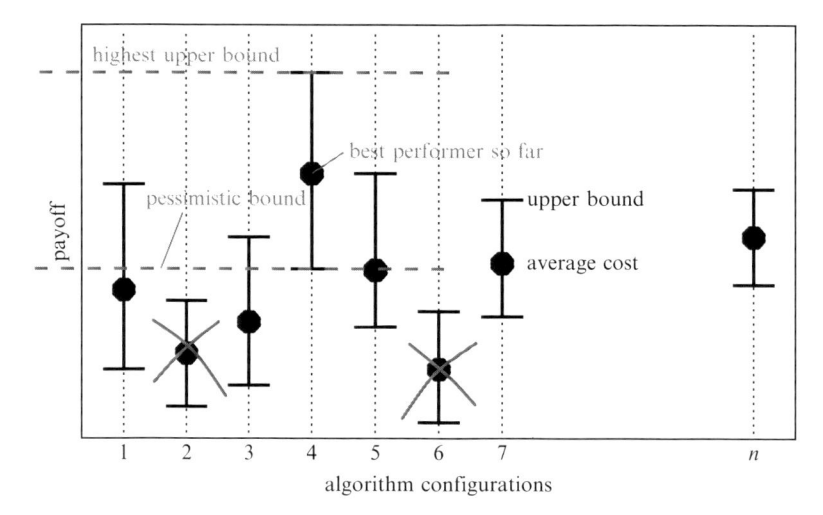

Fig. 10.6 Racing for off-line optimal configuration of metaheuristics. At each iteration, an estimate of the expected performance with error bars is available. Error bars are reduced when more tests are executed, and their values depend also on confidence parameter δ. In the figure, configurations 2 and 6 perform significantly worse than the best performer 4 and can be immediately eliminated from consideration: even if the real value of their performance is at the top of the error bar they cannot beat number 4

In [177], racing is used to select models in a supervised learning context, in particular for *lazy* or memory-based learning. Two methods are proposed for calculating error bars. One is based on Hoeffding's bound, which makes the only assumption of independence of the samples: the probability that the true error E_{true} being more than ε away from the estimate E_{est} is:

$$Prob(\|E_{\text{true}} - E_{\text{est}}\| > \varepsilon) < 2e^{\frac{-n\varepsilon^2}{B^2}}, \quad (10.6)$$

where B bound the largest possible error. In practice, this can be heuristically estimated as some multiple of the estimated standard deviation. Given the confidence parameter δ for the right-hand side of (10.6) (we want the probability of a large error to be less than δ), one easily solves for the error bar $\varepsilon(n, \delta)$:

$$\varepsilon(n, \delta) = \sqrt{\frac{B^2 \log(2/\delta)}{2n}}. \quad (10.7)$$

If the accuracy ε and the confidence δ are fixed, one can solve for the required number of samples n. The value $(1 - \delta)$ is the confidence in the bound for a single model during a single iteration, additional calculations provide a confidence $(1 - \Delta)$ of selecting the best candidate after the entire algorithm is terminated [177].

Tighter error bounds can be derived by making more assumptions about the statistical distribution. If the evaluation errors are normally distributed one can use

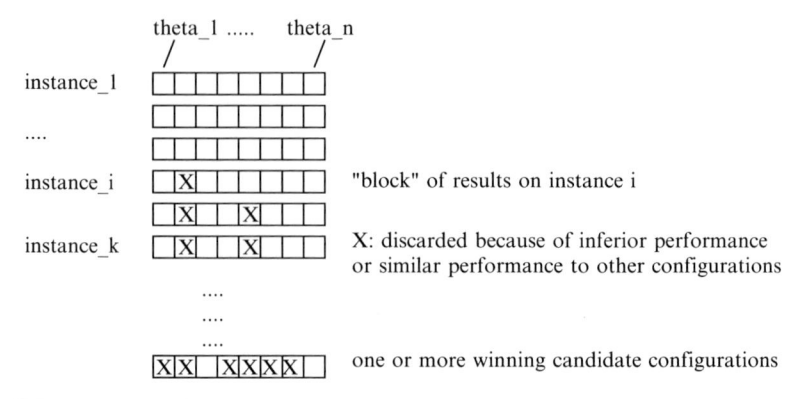

Fig. 10.7 Racing for off-line optimal configuration of metaheuristics. The most promising candidate algorithm configurations are identified *asap* so that these can be evaluated with a more precise estimate (more test instances). Each *block* corresponds to results of the various configurations on the same instance

Bayesian statistics, the second method proposed in [177]. One candidate model is eliminated if the probability that a second model has a better expected performance is above the usual confidence threshold:

$$\text{Prob}(E_{\text{true}}^{j} > E_{\text{true}}^{j'} \| e_j(1),...,e_j(n), e_{j'}(1),...,e_{j'}(n)) > 1-\delta. \qquad (10.8)$$

Additional methods for shrinking the intervals, as well as suggestions for using a statistical method known as blocking, are explained in [177]. Model selection in continuous space is considered in [88].

In [44] the focus is explicitly on metaheuristics configuration. Blocking through ranking is used in the F-RACE algorithm, based on the Friedman test, in addition to an aggregate test over all candidates performed before considering pairwise comparisons. Each block (Fig. 10.7) consists of the results obtained by the different candidate configurations θ_j on an additional instance i. From the results one gets a ranking R_{lj} of θ_j within block l, from the smallest to the largest, and $R_j = \sum_{l=1}^{k} R_{lj}$ the sum of the ranks over all instances. The Friedman test [74] considers the statistics T:

$$T = \frac{(n-1)\sum_{j=1}^{n}\left(R_j - \frac{k(n+1)}{2}\right)^2}{\sum_{l=1}^{k}\sum_{j=1}^{n}R_{lj}^2 - \frac{kn(n+1)^2}{4}}. \qquad (10.9)$$

Under the null hypothesis that the candidates are equivalent so that all possible rankings are equally likely, T is χ^2 distributed with $(n-1)$ degrees of freedom. If the observed t value exceeds the $(1-\delta)$ quantile of the distribution, the null hypothesis is rejected in favor of the hypothesis that at least one candidate tends to

10.4 Racing for Off-Line Configuration of Metaheuristics

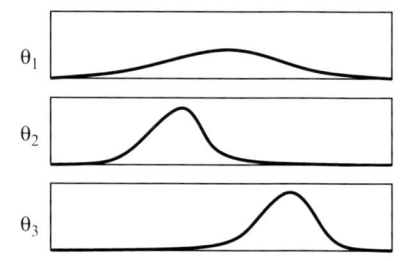

Fig. 10.8 Bayesian elimination of inferior models, from the posterior distribution of costs of the different models, one can eliminate the models that are inferior in a statistically significant manner, for example model θ_3 in the figure, in favor of model θ_2, while the situation is still undecided for model θ_1

perform better than at least another one. In this case one proceeds with a pairwise comparison of candidates. Configurations θ_j and θ_h are considered different if:

$$\frac{\|R_j - R_h\|}{\sqrt{\frac{2k(1-\frac{T}{k(n-1)})\left(\sum_{l=1}^{k}\sum_{j=1}^{n} R_{ij}^2 - \frac{kn(n+1)^2}{4}\right)}{(k-1)(n-1)}}} > t_{1-\delta/2}, \qquad (10.10)$$

where $t_{1-\delta/2}$ is the $(1-\delta/2)$ quantile of the Student's t distribution. In this case, the worse configuration is eliminated from further consideration.

Chapter 11
Teams of Interacting Solvers

> *No man is an island, entire of itself... any man's death diminishes me, because I am involved in mankind; and therefore never send to know for whom the bell tolls; it tolls for thee.*
> (John Donne)
>
> *A camel is a horse designed by committee.*
> (Sir Alexander Arnold Constantine Issigonis)

11.1 Complex Interaction and Coordination Schemes

Until now our discussion was mostly dedicated to single search schemes, generating a single trajectory while exploring the search space. An exception has been in the previous Chaps. 9 and 10, dedicated to portfolios and racing, where multiple searchers have been considered, either to improve the average behavior and reduce the risk or to select the best horses by feedback derived during an efficient parallel evaluation of more candidates. Let us now focus on more *complex interaction and coordination* schemes.

In this area, biological analogies derived from the behavior of different species abound. Elegant flocks of birds search for food or migrate in effective manners, herds of sheep get better guidance and protection than isolated members. Groups of hunting animals can often prevail over much bigger and powerful but isolated ones. Our species is particularly remarkable in this capability. After all, our culture and civilization has a lot to do with cooperation and coordination of intelligent human beings. Coming closer to our daily work, the history of science shows a steady advancement caused by the creative interplay of many individuals, both during their lifetimes and though the continuation of work by the future generations. We are all dwarfs standing on the shoulders of giants (*nanos gigantium humeris insidentes*, Bernard of Chartres).

Analogies from nature can be inspiring but also misleading when they are translated directly into procedures for efficient and effective problem solving and optimization. Let us consider a flock of birds or an ant colony searching for food. If an individual finds food, it makes perfect sense for the survival of the species to inform other members so that they can also get their share of nutrients. The analogy between food and good solutions of an optimization problems is not only far-fetched but quite simply wrong. If one searcher already found a good suboptimal solution, attracting other searchers in the same attraction basin around the locally optimal point only means wasting precious computational resources that could be spent by exploring different regions of the search space.

One encounters here the basic tradeoff between intensification and diversification. By the way, in the history of science, intensification has to do with research within a prevailing paradigm, while strong diversification has to do with a paradigm shift. According to a model proposed by Kuhn [165], when enough significant anomalies have accrued against a current *paradigm*, the scientific discipline is thrown into a state of crisis. During this crisis, new ideas are tried. Eventually, a new paradigm is formed, which gains its own new followers.

The adoption of a set of interacting search streams has a long history, not only when considering natural evolution, but also heuristics and learning machines. It is not surprising to find very similar ideas under different names, including ensembles, pools, agents, committees, multiple threads, mixture of experts, genetic algorithms, particle swarms, evolutionary strategies. The terms are not synonymous because each parish church has specific strong beliefs and true believers.

We tend to prefer the term *solver teams* to underline that an individual solver can be an arbitrarily intelligent agent capable of collecting information, developing models, exchanging the relevant part of the obtained information with his fellow searchers, and responding in a strategic manner to the information collected. Teamwork is the concept of people working together cooperatively, as in a sports team.

Why does it make sense to consider separately the team members from the team? After all one could design a complex entity where the boundary between the individual team members and the coordination strategy is fuzzy, a kind of Star Trek *borg hive* depicted as an amalgam of cybernetically enhanced humanoid drones of multiple species, organized as an interconnected collective with a hive mind and assimilating the biological and technological distinctiveness of other species to their own, in pursuit of perfection.

The answer lies in the simpler design of more advanced and effective search strategies and in the better possibility to *explain* the success of a particular team by separating its members' capabilities from the coordination strategy. Let us note that alternative points of view exist and are perfectly at home in the scientific arena: For example, one may be interested in explaining how and why a collection of very simple entities manages to solve problems not solvable by the individuals. For example, how simple ants manage to transport enormous weights by joining forces. The specific point of view demands that a *solver team* existence is not motivated by its sexiness or its correspondence to poetic biological analogies, but only by a demonstrated superiority with respect to the state of the art of the individual solvers.

11.2 Genetic Algorithms and Evolution Strategies

A rich source of inspiration for adopting a set of evolving candidate solutions is derived from the theory, natural evolution of the species, dating back to the original book by Charles Darwin [82] "On the Origin of Species by Means of Natural Selection, or the preservation of favored races in the struggle for life." It introduced the theory that populations evolve over the course of generations through a process

11.2 Genetic Algorithms and Evolution Strategies

of natural selection: Individuals more suited to the environment are more likely to survive and more likely to reproduce, leaving their inheritable traits to future generations. After the more recent discovery of the genes, the connection with optimization is as follows: each individual is a candidate solution described by its genetic content (genotype). The genotype is randomly changed by mutations, the suitability for the environment is described by a *fitness function*, which is related to the function to be optimized. Fitter individuals in the current population produce a larger offspring (new candidate solutions), whose genetic material is a recombination of the genetic material of their parents.

Seminal works include [16,101,134,214,226]. A complete presentation of different directions in this huge area is out of the scope of this chapter, let us concentrate on some basic ideas, and on the relationships between GA and intelligent search strategies.

To fix ideas, let us consider an optimization problem where the configuration is described by a binary string, mutation consists of randomly changing bit values with a fixed probability Π_{mute} and recombination consists of the so called *uniform crossover*: starting from two binary strings X and Y a third string Z is derived where each individual bit is copied from X with probability $1/2$, from Y otherwise.

The pseudo-code for a slightly more general version of a genetic algorithm, with an additional parameter for cross-over probability Π_{cross} is shown in Fig. 11.1, and illustrated in Fig. 11.2. After the generation of an initial population P, the algorithm iterates through a sequence of basic operations: First the fitness of each individual is computed and, if goals are met, the algorithm stops. Some individuals are chosen by a random selection process that favors elements with a high fitness function; a crossover is applied to randomly selected pairs to combine their features, then some individuals undergo a random mutation. The algorithm is then repeated on the new population.

Let the population of candidate solutions be a set of configurations scattered on the fitness surface. Such configurations explore their neighborhoods through the mutation mechanism: Usually the mutation probability is very low, so that in the above example, a small number of bits is changed. After this very primitive form of local search move, the population is substituted by a new one, where the better points have a larger probability to survive and a larger probability to generate offspring points. With a uniform cross-over, the offspring is generated though a kind of "linear interpolation" between the two parents. Because a real interpolation is excluded for binary values, this combination is obtained by the random mechanism described: If two bits at corresponding positions in X and Y have the same value, this value is maintained in Z – as it is in a linear interpolation of real-valued vectors – otherwise a way to define a point in between is to pick randomly from either X or Y if they are different.

There are at least three critical issues when adopting biological mechanisms for solving optimization problems: First, one should demonstrate that they are effective – not so easy because defining the function that biological organisms are optimizing is far from trivial. One risks a circular argument: survival of the fittest means that the fittest are the ones who survived. Second, one should demonstrate that they

Initialization — Compute a random population with M members $P = \{s^{(j)} \in S^l, j = 0, \ldots, M-1\}$, where each string is built by randomly choosing l symbols of S.

Repeat :

> **Evaluation** — Evaluate the fitness $f^{(i)} = f(s^{(i)})$; compute the rescaled fitness $\bar{f}^{(j)}$:
>
> $$f_{\min} = \min_{j=0,\ldots,M-1} f^{(j)}; \quad f_{\max} = \max_{j=0,\ldots,M-1} f^{(j)}; \quad \bar{f}^{(j)} = \frac{f^{(j)} - f_{\min}}{f_{\max} - f_{\min}}$$
>
> **Test** — If the population P contains one or more individuals achieving the optimization goal within the requested tolerance, stop the execution.
>
> **Stochastic selection** — Build a new population $Q = \{q^{(j)}, j = 0, \ldots, M-1\}$ such that the probability that an individual $q \in Q$ is member of P is given by $f(q)/\sum_{p \in P} f(p)$:
>
> **Reproduction** — Choose $N/2$ distinct pairs $(q^{(i)}, q^{(j)})$ using the N individuals of Q. For each pair build, with probability Π_{cross}, a new pair of offsprings mixing the parents' genes, otherwise copy the original genes:
>
> for $i \leftarrow 0, \ldots, (M-1)/2$
> > if RAND(1) < Π_{cross}
> > > for $j \leftarrow 0, \ldots, l-1$
> > > > if RAND(1) < .5
> > > > > $\bar{q}_j^{(2i)} \leftarrow q_j^{(2i)}; \bar{q}_j^{(2i+1)} \leftarrow q_j^{(2i+1)}$
> > > >
> > > > else
> > > > > $\bar{q}_j^{(2i)} \leftarrow q_j^{(2i+1)}; \bar{q}_j^{(2i+1)} \leftarrow q_j^{(2i)}$
> >
> > else
> > > $\bar{q}^{(2i)} \leftarrow q^{(2i)}; \bar{q}^{(2i+1)} \leftarrow q^{(2i+1)}$
>
> **Mutation** — In each new individual $\bar{q}^{(j)}$ change, with probability Π_{mute}, each gene $\bar{q}_i^{(j)}$ with a randomly chosen gene of S. Let us denote the new population Q'.
>
> **Replacement** — Replace the population P with the newly computed one Q'.

Fig. 11.1 Pseudocode for a popular version of GA

are efficient. Just consider how many generations were needed to adapt our species to the environment. Third, even assuming that GA are effective, one should ask whether natural evolution proceeds in a Darwinian way because it is intrinsically superior or because of hard biological constraints. For example, it is now believed that Lamarck was wrong in assuming that the experience of an individual could be passed to his descendants: the genes do not account for modifications caused by learning. But airplanes do not flap their wings and, in a similar manner, when a technological problem has to be solved, one is free to depart from the biological analogy and consider complete freedom to design the most effective method, as we will see in a moment. A second departure from the biological world is as follows: In biology, one is interested in the convergence of the entire population/species to a high fitness value, whereas in optimization one aims at having at least *one* high-fitness solutions *during* the search, not necessarily at the end, and one could not care less whether most individuals are far from the best when the search is terminated.

When using the stochastic local search language adopted in this book, the role of the mutation/selection and recombination operators cannot be explained in a

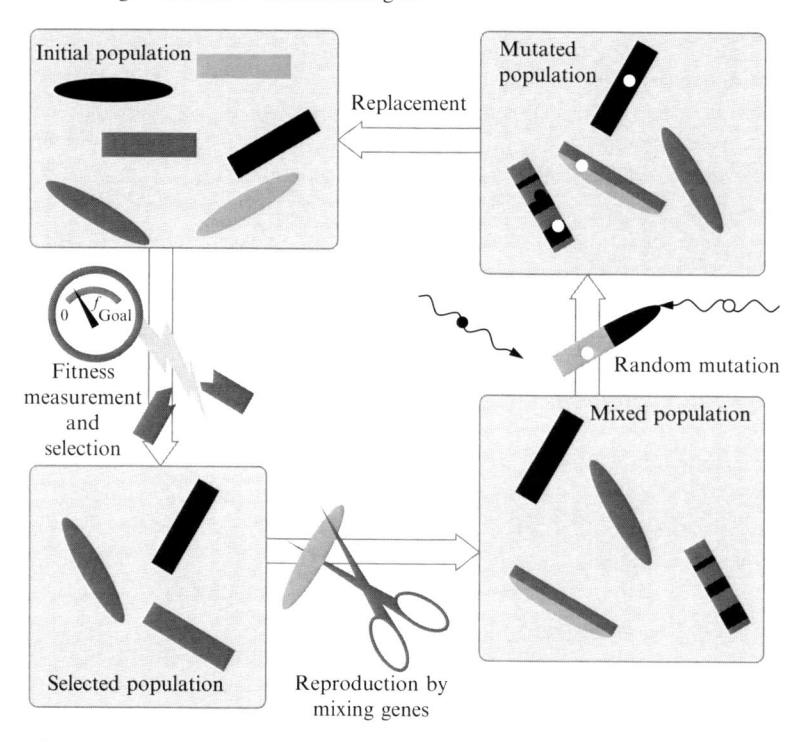

11.2 Genetic Algorithms and Evolution Strategies

Fig. 11.2 Representation of a genetic algorithm framework. Counter-clockwise from *top left*: starting from an initial population, stochastic selection is applied, then genes are mixed and random mutations occur to form a new population. When goals are met, the algorithm stops

clear-cut manner. When the mutation rate is small, the effect of the combined mutation and selection of the fittest can be interpreted as searching in the neighborhood of a current point, and accepting (selecting) the new point in a manner proportional to the novel fitness value. The behavior is of intensification/exploitation provided that the mutation rate is small, otherwise one ends up with a random restart, but not too small, otherwise one is stuck with the starting solution. The explanation of cross-over is more dubious. Let us consider uniform cross-over: If the two parents are very different, the distance between parents and offspring is large so that cross-over has the effect of moving the points rapidly on the fitness surface, while keeping the most stable bits constant and concentrating the exploration on the most undecided bits, the ones varying the most between members of the population. But if the similarity between parents is large, the cross-over will have little effect on the offspring, the final danger being that of a *premature convergence* of the population to a suboptimal point. The complexity inherent in explaining Darwinian metaphors for optimization makes one think whether more direct terms and definitions should be used [80] ("metaphors are not always rhetorically innocent").

At this point, you may wonder whether the term "team member" is justified in a basic GA algorithm. After all, each member is a very simple individual indeed, it comes to life through some randomized recombination of its parents and does a little exploration of its neighborhood. If it is lucky by encountering a better fitness value in the neighborhood, it has some probability to leave some of its genetic material to its offspring, otherwise it is terminated. No memory is kept of the individuals, only a collective form of history is kept through the population. Yes, "team member" *is* exaggerated. But now let us come back to the issue "airplanes do not flap their wings" to remember that as computer scientists and problem solvers our design freedom is limited only by our imagination.

We may imagine at least two different forms of *hybridized genetic algorithms*. The first observation is that, to deserve its name, a team member can execute a more directed and determined exploitation of its initial genetic content (its initial position). This is effected by considering the initial string as a *starting point* and by initiating a run of local search from this initial point, for example scouting for a local optimum. Lamarck can have his revenge here, now nobody prohibits substituting the initial individual with its ameliorated version after the local search. The term *memetic algorithms* [164, 191] has been introduced for models that combine the evolutionary adaptation of a population with individual learning within the lifetime of its members. The term derives from Dawkins' concept of a *meme*, which is a unit of cultural evolution that can exhibit local refinement [83].

Actually, there are two obvious ways in which individual learning can be integrated: First way consists of replacing the initial genotype with the better solution identified by local search (*Lamarckian evolution*), and second way can be of modifying the fitness by taking into account not the initial value but the final one obtained through local search. In other words, the fitness does not evaluate the initial state but the value of the "learning potential" of an individual, measured by the result obtained after the local search. This has the effect of changing the fitness landscape, while the resulting form of evolution is still Darwinian in nature. This and related forms of combinations of learning and evolution are known as the *Baldwin effect* [131, 267], see also the discussion in Sect. 7.4.1 about the Memory-Based Affine Shaker method.

11.3 Intelligent and Reactive Solver Teams

When one departs from biology and enters optimization technology, the appetite for more effective and efficient basic solvers can be satisfied by adopting more complex techniques, embodying elements of metaoptimization and self-tuning.

When one considers the relevant issues in designing solver teams, one encounters striking analogies with sociological and behavioral theories related to human teams, with a similar presence of apparently contradictory conclusions. Let us mention some of the basic issues from a qualitative and analogical point of view, deferring more specific algorithmic implementations in the next sections.

Individual quality of the team members: The basic issue here is related to the relationship between the quality of the members and the collective quality of the

11.3 Intelligent and Reactive Solver Teams

team. If the team members are poor performers, one should not expect an exceptional team. Nonetheless, it is of scientific and cultural interest to assess the potential for problem solving through the interaction of a multitude of simple members. An inspiring book [246] deals with the *wisdom of crowds*: "why the many are smarter than the few and how collective wisdom shapes business, economies, societies, and nations." Of course, acid comments go back ages, like in Nietzsche's citation "I do not believe in the collective wisdom of individual ignorance." Adapting a conclusion from the review [215], in matters for which true expertise can be identified, one would much rather rely on the best judgments of the most knowledgeable specialists than a crowd of laymen. But in matters for which no expertise or training is genuinely involved, in dealing with fields of study whose principles are ambiguous, contentious, and rarely testable, ... then yes, there is sense to polling a group of people." In optimization, one expects a much bigger effectiveness by starting from the most competitive single-thread techniques, but some dangers lurk depending on how the team is integrated and information is shared and used.

Diversity of the team members: If all solvers act in same manner the advantage is lost (see lower panes of Fig. 11.3). Diversity can be obtained in many possible ways, for example by using different solvers, or identical solvers but with different random initializations. In some cases, the effects of an initial randomization is propagated throughout a search process that shows a sensitive dependence on initial conditions that is characteristic for chaotic processes [251].

Diversity means that, in some cases, combining simpler and inferior performers with more effective ones can increase the overall performance and/or improve robustness. By the way, diversity is also crucial for ensembles of learning machines [25, 254].

Information sharing and cooperation strategy: When designing a *solver team*, one must decide about the way in which information collected by the various solvers is shared and used to modify the individual server decisions. An extreme design principle consists of complete independence and periodic reporting of the best solution to a coordinator. Simplicity and robustness make this extreme solution the first to try: More complex interaction schemes should always be compared against this baseline, see for example [26] where independent parallel walks of tabu search are considered.

More complex sharing schemes involve periodic collection of the best-so-far values and configurations, the current configurations, more complex summaries of the entire search history of the individual solvers, for example the list of local minima encountered.

After the information is shared, a decision process modifies the individual future search in a strategic manner. Here complexities and open research issues arise. Let us consider a simple case: If a solver is informed by a team member about a new best value obtained corresponding to a configuration that is far from the current search region, is it better for it to move to the new and promising area or to keep visiting the current region? See also Fig. 11.3.

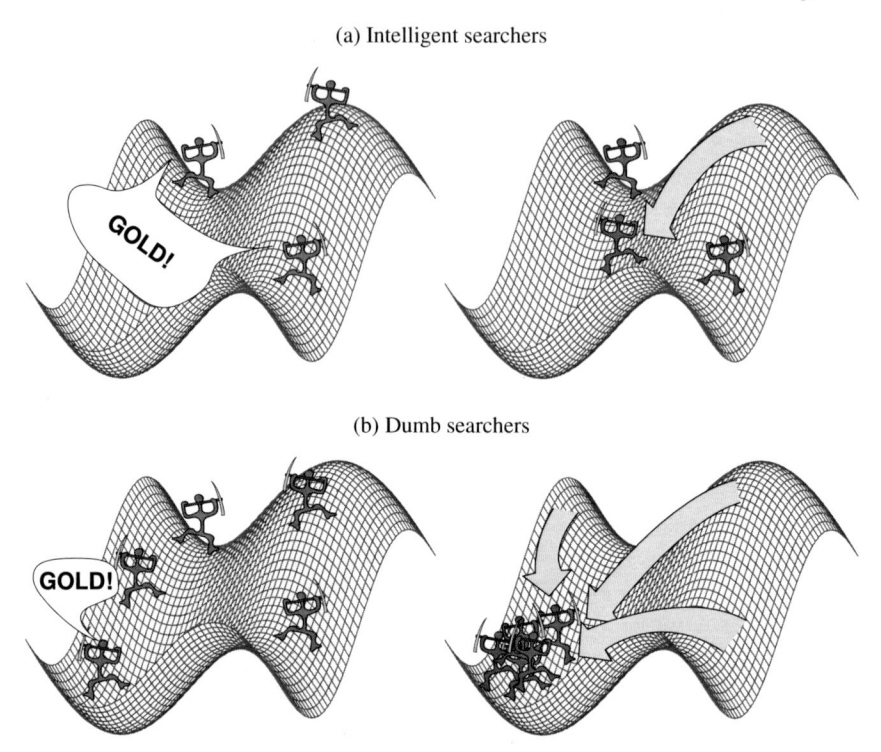

Fig. 11.3 Opportunities and pitfalls of teams of searchers: an intelligent searcher may decide to move between two successful ones (*top panes*), while dumb searchers (*bottom*) swarm to the same place, so that diversification can be lost. If you were a 49er, which model would you choose (and, besides, would you scream your findings)?

An example of the dangerous effects of interaction in the social arena is called *groupthink* [148], a mode of thinking that people engage in when they are deeply involved in a cohesive group, when the members' strivings for unanimity override their motivation to realistically appraise alternative courses of action. *Design by committee* is a second term referring to the poor results obtained by a group, particularly in the presence of poor and incompetent leadership.

An example of a pragmatic use of memory in cooperation is [210], where experiments highlight that "memory is useful but its use from the very beginning of the search is not recommended."

Centralized vs. distributed management schemes: This issue is in part related to computing hardware and communication networks, in part related to the software abstraction adopted for programming. The centralized schemes, often related to some form of synchronous communication, see a central coordinator acting in a *master-slave* relationship with respect to the individual solvers. The team members are given instructions (for example initialization points, parameters, detailed strategies) by the coordinator and periodically report about search progress and other parameters. The opposite point consists of distributed computation by peers, which periodically and often asynchronously exchange information and autonomously decide about their future steps. Gossiping optimization schemes fall in this category. The design alternatives are related to efficiency but also simplicity of programming and understanding. For example, a synchronous parallel machine may be handled more efficiently through a central coordination, while a collection of computers distributed in the world, prone to disconnections and connected by internet, may find a more natural coordination scheme by *gossiping*. Reviews of parallel strategies for local search and metaheuristics are presented in [257] (see for example the *multiple walks* parallelism), in [115] ("controlled pool maintenance") and [78].

Reactive vs. nonreactive schemes: Last but not least comes the issue of learning on the job and fine self-tuning of algorithm parameters. In addition to the adaptation of individual parameters based on individual search history, which is by now familiar to the reader, new possibilities for a reactive adaptation arise by considering the search history of other team members and the overall coordination scheme. As examples, adaptation in evolutionary computation is surveyed in [90, 130]. Adaptation can act on the representation of the individuals, the evaluation function, the variation, selection and replacement operators and their probabilities, the population (size, topology, etc.). In their taxonomy, parameter tuning coincides with our "off-line tuning," while parameter control coincides with "on-line tuning," adaptive parameter control has a reactive flavor, while self-adaptive parameter control means that one want to use the same golden hammer (GA) for all nails, including metaoptimization (the parameters are encoded into chromosomes). Strategic design embodying intelligence more than randomization is also advocated in the *scatter search and path relinking approach* [110, 111]. Scatter search is a general framework to maintain a reference set of solutions, determined by their function values but by their level of diversity, and to build new ones by "linearly interpolating and extrapolating" between subsets of these solutions. Of course, interpolation must be interpreted to work in a discrete setting (the above considered uniform cross-over is a form of interpolation) and adapted to the problem structure. *Path relinking* generalizes the concept: Instead of creating a new solution from a set of two parents, an entire path between them is created by starting from one extreme and progressively modifying the solution to reduce the distance from the other point. The approach is strongly pragmatic, alternatives are tried and judged in a manner depending on the final results and not on the adherence to biological and evolutionary principles, as it should always be the case if the final interest is effective problem-solving.

11.4 An Example: Gossiping Optimization

Let us consider the following scenario: a set of intelligent searchers is spread on a number of computers, possibly throughout the world. While every searcher executes a local search heuristic, it takes advantage from the occasional injection of new information coming from its partners on other machines.

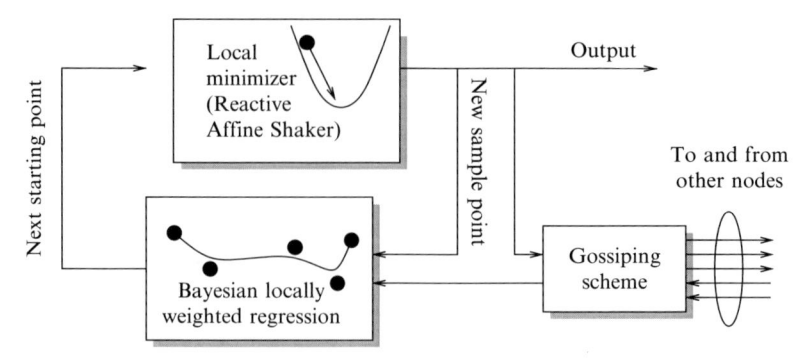

Fig. 11.4 The distributed setting (compare with Fig. 7.14)

For instance, let us consider the field of continuous optimization, for which the memory-based MRAS heuristic has been introduced in Sect. 7.4.1. A fast local minimizer, the reactive affine shaker, interacts with a model of the search space by feeding it with new data about the search and retrieving suggestions about the best starting point for a new run.

While the MRAS algorithm has been devised as a sequential heuristic [58], it can be extended to a distributed one as described in Fig. 11.4: a gossiping component, described below, is used to communicate model information to other nodes, and to feed information coming from other nodes to the model.

Sharing information among nodes is a delicate issue. The algorithm aims at function optimization, so it should spend most of its time doing actual optimization, not just broadcasting information to other nodes. Let us now discuss some communications issues arising in this context.

11.4.1 Epidemic Communication for Optimization

The use of parallel and distributed computing for solving complex optimization tasks has been investigated extensively in the last decades [26, 249]. Most works assume the availability of either a dedicated parallel computing facility or of a specialized clusters of networked machines that are coordinated in a centralized fashion (master-slave, coordinator-cohort, etc.). Although these approaches simplify management, they have limitations with respect to scalability and robustness and require a dedicated investment in hardware.

Recently, the peer-to-peer (P2P) paradigm for distributed computing has demonstrated that networked applications can scale far beyond the limits of traditional distributed systems without sacrificing efficiency and robustness. The application of P2P is not limited to content distribution, as most popular accounts suggest: it also covers scientific purposes related to solving massive computation problems in the absence of a dedicated infrastructure.

11.4 An Example: Gossiping Optimization

On the one hand, a well known problem with P2P systems is their high level of dynamism: nodes join and leave the system continuously, in many cases unexpectedly and without following any "exit protocol." This phenomenon, called *churn*, together with the large number of computational nodes are the two most prominent research challenges posed by P2P systems: No node has an up-to-date knowledge of the entire system, and the maintenance of consistent distributed information may as well be impossible.

On the other hand, a clear advantage of using a P2P architecture is the exploitation of unused computational resources, such as personal desktop machines, volunteered by people who keep using their computers while participating to a shared optimization effort. The systems based on a central coordinator repeat a simple loop: Every involved machine receives from a central server a subset of the search space (samples and parameter intervals), performs an exhaustive coverage of the subset and reports the results, receiving another search subset.

More distributed schemes originated in the context of databases [85], where *epidemic* protocols have been able to deal with the high levels of unpredictability associated with P2P systems. Apart from the original goal of information dissemination (messages are "broadcasted" through random exchanges between nodes), epidemic protocols are now used to solve several different problems, from membership and topology management to resource sharing.

We focus our attention onto stochastic local search schemes based on memory, where little or no information about the function to be optimized is available at beginning of the search. In this context, the knowledge acquired from function evaluations at different input points during the search can be mined to build models so that the future steps of the search process can be optimized. An example is the online adaptive self-tuning of parameters while solving a specific instance proposed by Reactive Search. Recent developments of interest consider the integration of multiple techniques and the feedback obtained by preliminary phases of the execution for a more efficient allocation of the future effort, see for example the a-teams scheme in [240], the portfolios proposals [113, 142], the racing schemes [68, 242], dynamic restart policies [157].

In the P2P scenario, the crucial issues and tradeoffs to be considered when designing distributed optimization strategies are:

– Coordination and interaction: One has a choice of possibilities ranging from independent search processes reporting the end results, to fully coordinated "teams" of searchers exchanging new information after each step of the search process.
– Synchronization: In a peer-to-peer environment, the synchronization must be very loose to avoid wasting computational cycles while waiting for synchronization events.
– Type and amount of exchanged information: It ranges from the periodic exchange of current configurations and related function values, see for example particle swarm [70] and genetic algorithms, to the exchange of more extensive data about past evaluations, possibly condensed into local heuristic models of the function [58].

– Frequency of gossiping, convergence issues: We consider a simple basic interaction where a node picks a random neighbor, exchanges some information, and updates its internal state (memory). The spreading of the information depends both on the gossiping frequency and on the interconnection topology. Tradeoffs between a more rapid information exchange and a more rapid advancement of each individual search process are of interest.
– Effects of delays on "distributed snapshots": Because of communication times, congestion and possible temporary disconnections, the received information originated from a node may not reflect accurately the current state, so that decisions are made in a suboptimal manner.

The distributed realization of a gossiping optimization scheme ranges between two extremes:

- *Independent execution of stochastic processes.* Some global optimization algorithms are stochastic by nature; in particular, the first evaluation is not driven by prior information, so the earliest stages of the search often require some random decision. Different runs of the same algorithm can evolve in a very different way, so that the parallel independent execution of identical algorithms with different random seeds permits to explore the tail of the outcome distribution toward lower values.
- *Complete synchronization and sharing of information.* Some optimization algorithms can be modeled as parallel processes sitting in a multiprocessor machine supporting shared data structures. Processes can be coordinated in such a way that every single step of each process, i.e., decision on the next point to evaluate, is performed while considering information about all processes. An example is the Particle Swarm algorithm, where the optimum is searched along multiple trajectories whose evolution depends both on personal and global history through a very simple rule. The particle swarm method is usually implemented as a single process operating on an array of evaluation points, but its parallelization is straightforward, provided that the cost of sharing global information does not overcome the advantage of having many function evaluations performed simultaneously.

Between the two extremal cases, a wide spectrum of algorithms can be designed to perform individual searches with some form of loose coordination. An example is the "GOSH!" paradigm (Gossiping Optimization Search Heuristics) proposed in [43]. In this proposal, to distribute the Memory-Based Affine Shaker (MRAS) algorithm, every node maintains its own history and uses it to model the function landscape and locate the best suitable starting point. Occasionally, pairs of nodes communicate and share relevant information about their history to build a better common model. The frequency and randomness of information exchanges are the crucial parameters in this case.

Chapter 12
Metrics, Landscapes, and Features

La filosofia è scritta in questo grandissimo libro che continuamente ci sta aperto innanzi a gli occhi (io dico l'universo), ma non si può intendere se prima non s'impara a intender la lingua, e conoscer i caratteri, né quali è scritto. Egli è scritto in lingua matematica, e i caratteri son triangoli, cerchi, ed altre figure geometriche, senza i quali mezzi impossibile a intenderne umanamente parola; senza questi un aggirarsi vanamente per un oscuro laberinto.

Philosophy is written in this grand book (I mean the universe) which stands continually open to our gaze, but it cannot be understood unless one first learns to comprehend the language in which it is written. It is written in the language of mathematics, and its characters are triangles, circles, and other geometric figures, without which it is humanly impossible to understand a single word of it; without these, one is wandering about in a dark labyrinth.
(Galileo, Il Saggiatore, 1623)

Measure what is measurable, and make measurable what is not so.
(Galileo Galilei)

12.1 How to Measure and Model Problem Difficulty

When using learning strategies in the area of heuristics, finding appropriate features to measure and appropriate metrics is a precious guide for the design of effective strategies and an indispensable factor if one aims at explanation and understanding.

Let us consider some challenging questions:

- How can one predict *the future evolution of an algorithm*? For example, the running time to completion, the probability of finding a solution within given time bounds, etc.
- How can one determine which is *the most effective heuristic* for a given problem, or for a specific instance?
- How can one identify the *intrinsically more difficult problems or instances* for a given search technique?

Although a complete review of the literature in this area is beyond the scope of this book, in this chapter we mention some interesting research issues related to measuring problem difficulty, measuring individual algorithm components, and selecting them through a diversification and bias metric.

Let us consider the issue of understanding why a problem is more difficult to solve for a stochastic local search method. One aims at discovering relationships between problem characteristics and problem difficulty. Because the focus is on local search methods, one would like to characterize statistical properties of the solution *landscape* leading to a more difficult exploration.

The effectiveness of a stochastic local search method is determined by how *microscopic local decisions* made at each search step interact to determine the *macroscopic global behavior* of the system, in particular, how the function value f depends on the input configuration. Statistical mechanics has been very successful in the past at relating local and global behaviors of systems [132], for example starting from the molecule–molecule interaction to derive macroscopic quantities like pressure and temperature. Statistical mechanics builds upon statistics, by identifying appropriate statistical *ensembles* (configurations with their probabilities of occurrence) and deriving typical global behaviors of the ensemble members. When the numbers are large, the variance in the behavior is very small, so that most members of the ensemble will behave in a similar way. On the one hand, if one has two communicating containers of one liter and a gas with five flying molecules, the probability to find all molecules in one container is not negligible. On the other hand, the probability to observe 51% of the molecules in one container is very close to zero if the containers are filled with air at normal pressure: even if the individual motion is very complex, the macroscopic behavior will produce a 50% subdivision with a very small and hardly measurable random deviation (the molecule count is left as an exercise to the reader).

Unfortunately, the situation for combinatorial search problems is much more complicated than the situations for physics-related problems, so that the precision of theoretical results is more limited. Nonetheless, a growing body of literature exists, which sheds light onto different aspects of combinatorial problems and permits *a level of understanding and explanation which goes beyond the simple empirical models* derived from massive experimentation.

12.2 Phase Transitions in Combinatorial Problems

Models inspired by statistical mechanics have been proposed for some well-known combinatorial problems. For example, an extensive review of models applied to constraint satisfaction problems, in particular the graph coloring problem, is present in [132]. The SAT problem, in particular the 3-SAT, has been the playground for many investigations, see for example [69, 79, 206, 234].

Phase-transitions have been identified as a mechanism to study and explain problem difficulty. A phase-transition in a physical system is characterized by the abrupt change of its macroscopic properties at certain values of the defining parameters. For example, consider the transitions from ice to water to steam at specific values of temperature and pressure. Phenomena analogous to phase-transitions have been studied for random graphs [48, 92]: as a function of the average node degree, some macroscopic property like connectivity change in a very rapid manner. The work in [141] predicts that large-scale artificial intelligence systems and cognitive models will undergo sudden phase transitions from disjointed parts into coherent structures as their topological connectivity increases beyond a critical value. Citing from the paper: "this phenomenon, analogous to phase transitions in nature, provides a new paradigm with which to analyze the behavior of large-scale computation and determine its generic features."

Constraint satisfaction and SAT phase-transitions have been widely analyzed, for a few references see [61, 162, 186, 206, 232, 236]. A clear introduction to phase-transitions and the search problem is present in [133]. A surprising result is that hard problem instances are concentrated near the same parameter values for a wide variety of common search heuristics. This location also corresponds to a *transition between solvable and unsolvable instances*. For example, when the control parameter that is changed is the number of clauses in SAT instances, different schemes such as complete backtracking and local search show very long computing times in the same transition region.

For backtracking, this is due to a competition between two factors: (1) number of solutions and (2) facility of pruning many subtrees. A small number of clauses (*under-constrained* problem) implies many solutions, it is easy to find one of them. At the other extreme, a large number of clauses (*over-constrained* problem) implies that any tentative solution is quickly ruled out (pruned from the tree), it is fast to rule out all possibilities and conclude with no solution. The *critically constrained* instances in between are the hardest ones.

For local search one has to be careful. The method is not complete and one must limit the experimentation to solvable instances. One may naïvely expect that the search becomes harder with a smaller *number of solutions* but the situation is not so simple. At the limit, if only one solution is available but the attraction basin is very large, local search will easily find it. Not only the number of solutions but also the number and depth of *suboptimal local minima* play a role. A large number of deep local minima is causing a waste of search time in a similar way to tentative solutions in backtracking, which fail only after descending very deeply in the search tree. Intuition helps, for a growing body of experimental research see for example [69] for results on CSP and SAT, [79] for experimental results on the crossover point in random 3-SAT.

In addition to being of high scientific interest, identifying zones where the most difficult problems are is very relevant for *generating difficult instances to challenge algorithms*. As strange as it may sound at the beginning, it is not so easy to identify difficult instances for NP-hard problem, see for example [232] for generating hard Satisfiability problems. Let us remember that the computational complexity classes are defined through a *worst-case* analysis: in practice the worst cases may be very difficult to encounter or to generate.

12.3 Empirical Models for Fitness Surfaces

More empirical *descriptive cost models* of problem difficulty aim at identifying measurable *instance characteristics (features) influencing the search cost*. A good descriptive model should account for a significant portion of the variance in search cost.

The work [69] demonstrates that the logarithm of the number of optimal solutions accounts for a large portion of the variability in local search cost for the SAT problem. The papers [206] and [234] study the distribution of SAT solutions and demonstrate that the size of the *backbone* (the set of Boolean variables that have the same value in all optimal solutions) is positively correlated to the solution cost. The contribution [264] considers the job shop scheduling problem (JSP) and demonstrates experimentally that the mean distance between random local minima and the nearest optimal solution is highly correlated with the cost of solving the problem to optimality (a simple version of tabu search is used in the tests).

The performance of search algorithms depends on the features of the search space. In particular, a useful measure of variability of the search landscape is given by the correlation between the values of the objective function f over all pairs of configurations at distance d, where the definition of distance to be used depends on the nature of the solving technique. For instance, if local search is being employed, the distance between two configurations X and X' is measured as the minimum number of local steps to be taken to transform one into the other. Let us call it the *Landscape Correlation Function* [265]:

$$R(d) = \frac{E_{\text{dist}(X,X')=d}[(f(X)-\mu)(f(X')-\mu)]}{\sigma^2}, \quad (12.1)$$

where $\mu = E[X]$ and $\sigma^2 = \text{Var}[X]$. This measure captures the idea of *ruggedness* of a surface: a low correlation function implies high statistical independence between points at distance d. Although it is expectable that for large values of d the correlation $R(d)$ goes to zero (unless the search landscape is very smooth), the value of $R(1)$ can be meaningful.

Intuitively, $R(1)$ can tell us whether a local move from the current configuration changes the f value in a manner that is significantly different from a random restart. On the one hand, $R(1) \approx 0$ means that on the average there is little correlation between the objective value at a given configuration and the value of its neighbors. This can be an indication of a poor choice of the neighborhood structure, or that the problem is particularly hard for methods based on local search. On the other hand, $R(1) \approx 1$ is a clear indication that the neighborhood structure produces a *smooth* fitness surface.

Computing (12.1) for large search spaces can be difficult. A common estimation technique uses random walks. In particular, Big-Valley models [47] (a.k.a. *massif central* models) have been considered to explain the success of local search, and the preference for continuing from a given local optimum instead of restarting from scratch. These models measure the *autocorrelation of the time-series of f values produced by a random walk*. The autocorrelation function (ACF) of a random process describes the correlation between the process at different points in time. Let $X^{(t)}$ be the search configuration at time t. If $f(X^{(t)})$ has mean μ and variance σ^2 then the ACF can be defined as

$$R'(d) = \frac{E_t[(f(X^{(t)})-\mu)(f(X^{(t+d)})-\mu)]}{\sigma^2}. \quad (12.2)$$

12.3 Empirical Models for Fitness Surfaces

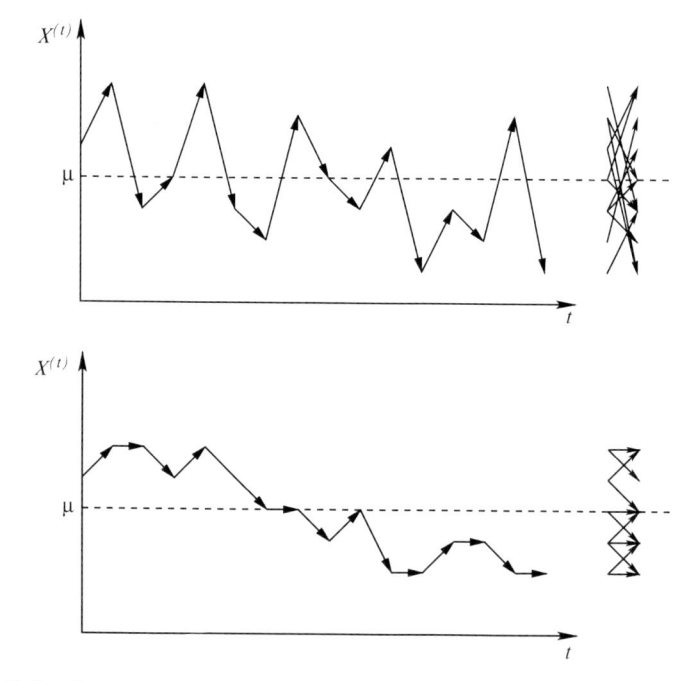

Fig. 12.1 Estimating autocorrelation on a rugged (*top*) and a smooth (*bottom*) landscape by means of a random walk. Vectors between fitness values at subsequent steps are shown on the *right*; in particular, $R'(1)$ is determined as the correlation between the endpoints of these *arrows* with respect to the mean value (*dashed line*)

One necessary assumption for (12.2) to adequately estimate (12.1) is that the random walk is representative of the whole landscape. This assumption is reasonable provided that the landscape is uniform, in particular the result should not depend on the initial configuration of the random walk.

Figure 12.1 provides a pictorial view of autocorrelation estimation by means of a random walk. In particular, the right side of each plot shows how subsequent moves are correlated: every arrow shows one step, so that correlation is computed between the head and the tail of each arrow with respect to the mean value (dashed line). It is apparent that the bottom walk represents a smoother landscape, and this translates to correlated arrow endpoints.

As noted earlier, it is expected that both $R(d)$ and $R'(d)$ become smaller and smaller as long as d increases: points separated by a small path are more correlated than separated ones. Empirical measurements show that $R'(d)$ often follows an exponential decay law:

$$R'(d) \approx e^{-\frac{d}{\lambda}}. \tag{12.3}$$

The value of λ that best approximates the $R'(\cdot)$ sequence is called the *correlation length*, and it is a measure of the range over which fluctuations of f in one region of space are correlated. An approximation of λ can be obtained by solving (12.3) when $d = 1$:

$$\lambda \approx -\frac{1}{\ln R'(1)}.$$

Clearly, we are assuming that correlation between nearby configurations is positive: otherwise, no significant approximation can be obtained and the whole correlation analysis loses its meaning.

Equation (12.3) defines the correlation length as the distance where autocorrelation decreases by a factor of e, so that the definition is somewhat arbitrary. Moreover, for many problems the correlation length is a function of the problem's size and it does not explain the variance in computational cost among instances of the same size [264]. However, it is possible to normalize it with respect to some measure of the problem's instance (e.g., number of dimensions, size of the neighborhood) to make it a useful tool for comparisons.

An example of fitness landscape analysis for the Quadratic Assignment Problem is presented in [183], where autocorrelation analysis and fitness-distance correlation analysis [154] are adopted for choosing suitable evolutionary operators.

The *fitness distance correlation* (FDC) is based on measuring the Hamming distances between sets of bit-strings and the global optimum, and comparing with their fitness. Large negative correlations between distance and fitness are taken to be indicators that the system is easy to optimize for GA. Large positive correlations indicate the problem is *misleading*, and selection will guide the population away from the global maximum. Near-zero correlations indicate that the GA does not have guidance toward or away from the optimum, and the search is close to random search. However, when the correlation coefficient is near zero, the FDC measure is simplistic and it requires a closer examination of the relation between fitness and distance, through the use of scatter plots of fitness versus distance. Of course, the technique can be generalized to deal with different distance measures. An instructive counterexample is presented in [6].

12.3.1 Tunable Landscapes

To develop intuition on the correlation structure of complex fitness landscapes, and to provide a parametric and tunable landscape for the generation of problem instances, let us consider the *NK landscape model* [156] proposed in the field of computational biology. *NK* landscapes try to capture the intuition that every coordinate in the search space contributes to the overall fitness in a complex way, often by enabling or disabling the contribution of other variables in a process that is known in biology as *epistasis*.

An NK system is defined by N binary components (e.g., bits in the configuration string) and by the number K of other components that interact with each component. In other words, each bit identifies a $(K+1)$-bit interaction group. A uniform

12.3 Empirical Models for Fitness Surfaces

random-distributed mapping from $\{0,1\}^{K+1}$ to $[0,1]$ determines the contribution of each interaction group to the total fitness, which is computed as the average of all contributions.

In mathematical terms, let the configuration space be the N-bit strings $\{0,1\}^N$. The current configuration is $S = s_1 \ldots s_N$, $s_i \in \{0,1\}$. For every position $i = 1, \ldots, N$, let us define its interaction group as a set of $K+1$ different indices $G_i = (g_{i0}, g_{i1}, \ldots, g_{iK})$ so that $g_{i0} = i$, $g_{ij} \in \{1, \ldots, N\}$ and if $l \neq m$ then $g_{il} \neq g_{im}$. Let $w : \{0,1\}^{K+1} \to [0,1]$, defined by random uniform distribution, represent the contribution of each interaction group. Then

$$f(S) = \frac{1}{N} \sum_{i=1}^{N} w(s_{g_{i0}} s_{g_{i1}} \ldots s_{g_{iK}}).$$

Figure 12.2 shows how every bit contributes to the fitness value by means of its epistatic interaction group as defined before. For small values of K the contribution function w can be encoded as a (2^{K+1})-entry lookup table.

Intuitively, parameter K controls the so-called "ruggedness" of the landscape: $K = 0$ means that no bitwise interaction is present, so that every bit independently contributes to the overall fitness, leading to an embarrassingly simple optimization task indeed. But then, if $K = N - 1$ then changing a single bit modifies all contributions to the overall fitness value: the ruggedness of the surface is maximized.

Because of their parametrically controlled properties, NK systems have been used as problem generators in the study of combinatorial optimization algorithms [126, 155, 237].

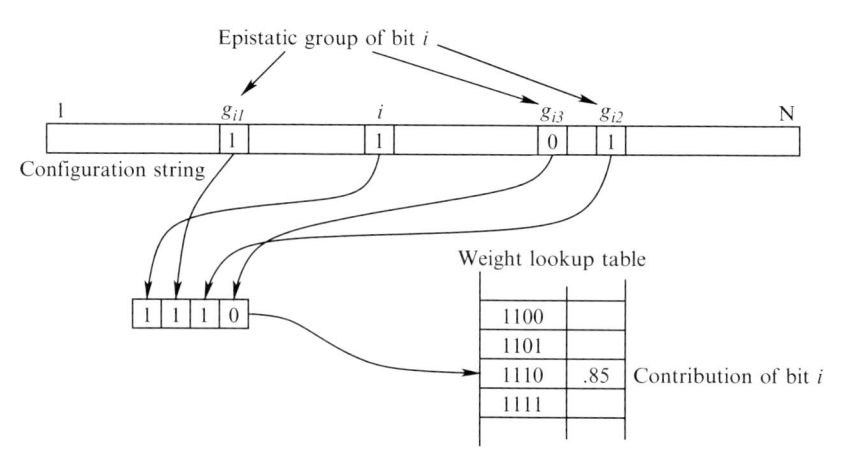

Fig. 12.2 Epistatic contribution of bit i in an NK model for $K = 3$

12.4 Measuring Local Search Components: Diversification and Bias

To ensure progress in algorithmic research, it is not sufficient to have a horse-race of different algorithms on a set of instances and declare winners and losers. Actually, very little information can be obtained by these kinds of comparisons. In fact, if the number of instances for the benchmark is limited and if sufficient time is given to an intelligent researcher (...and very motivated to get publication!) be sure that some that promising results will be finally obtained, via a careful tuning of algorithm parameters.

A better method is to design a *generator of random instances* so that it can produce instances used during the development and tuning phase, while a different set of instances extracted from the same generator is used for the final test. This method mitigates the effect of "intelligent tuning done by the researcher on a finite set of instances," and it can determine a winner in a fairer horse-race, but still does not explain *why* a method is better than another one. Explaining *why* is related to the *generality and prediction power* of the model. If one is capable of predicting the performance of a technique on a problem (or on a single instance) – of course before the run is finished, predicting the past is always easy! – then he takes some steps toward understanding.

This exercise takes different forms depending on what one is predicting, what are the starting data, what is the computational effort spent on the prediction, etc. To make some examples, the work in [28] dedicated to solving the MAX-SAT problem with nonoblivious local search aims at *relating the final performance to measures obtained after short runs of a method*. In particular, the average f value (*bias*) and the average speed in Hamming distance from a starting configuration (*diversification*) is monitored and related to the final algorithm performance.

Let us focus onto local-search based heuristics: it is well known that the basic compromise to be reached is that between *diversification* and *bias*. Given the obvious fact that only a negligible fraction of the admissible points can be visited for a nontrivial task, the search trajectory $X^{(t)}$ should be generated to visit preferentially points with large f values (*bias*) and to avoid the confinement of the search in a limited and localized portion of the search space (*diversification*). The two requirements are conflicting: as an extreme example, random search is optimal for diversification but not for bias. Diversification can be associated with different metrics. Here we adopt the *Hamming distance* as a measure of the distance between points along the search trajectory. The Hamming distance $H(X,Y)$ between two binary strings X and Y is given by the number of bits that are different.

The investigation in [28] follows this scheme:

- After selecting the metric (diversification is measured with the Hamming distance and bias with mean f values visited), the diversification of simple *random walk* is analyzed to provide a basic system against which more complex components are evaluated.

12.4 Measuring Local Search Components: Diversification and Bias

- The diversification-bias metrics (D–B plots) of different basic components are investigated and a conjecture is formulated that the best components for a given problem are the *maximal elements* in the diversification-bias (D–B) plane for a suitable partial ordering (Sect. 12.4.2).
- The conjecture is validated by a competitive analysis of the components on a benchmark.

Let us now consider the *diversification properties of Random Walk*. Random Walk generates a Markov chain by selecting at each iteration a random move, with uniform probability:

$$X^{(t+1)} = \mu_{r(t)} X^{(t)} \quad \text{where} \quad r(t) = \text{RAND}\{1,\ldots,n\}$$

Without loss of generality, let us assume that the search starts from the zero string: $X^{(0)} = (0,0,\ldots,0)$. In this case the Hamming distance at iteration t is:

$$H(X^{(t)}, X^{(0)}) = \sum_{i=1}^{n} x_i^{(t)}$$

and therefore the expected value of the Hamming distance at time t, defined as $\widehat{H}^{(t)} = \widehat{H}(X^{(t)}, X^{(0)})$, is:

$$\widehat{H}^{(t)} = \sum_{i=1}^{n} \widehat{x}_i^{(t)} = n\,\widehat{x}^{(t)}. \tag{12.4}$$

The equation for $\widehat{x}^{(t)}$, the probability that a bit is equal to 1 at iteration t, is derived by considering the two possible events that (a) the bit remains equal to 1 and (b) the bit is set to 1. In detail, after defining as $p = 1/n$ the probability that a given bit is changed at iteration t, one obtains:

$$\widehat{x}^{(t+1)} = \widehat{x}^{(t)}(1-p) + (1-\widehat{x}^{(t)})\,p = \widehat{x}^{(t)} + p(1 - 2\widehat{x}^{(t)}). \tag{12.5}$$

It is straightforward to derive the following theorem:

Theorem 1. *If $n > 2$ (and therefore $0 < p < \frac{1}{2}$) the difference equation (12.5) for the evolution of the probability $\widehat{x}^{(t)}$ that a bit is equal to one at iteration t, with initial value $\widehat{x}^{(0)} = 0$, is solved for t integer, $t \geq 0$ by:*

$$\widehat{x}^{(t)} = \frac{1 - (1-2p)^t}{2}. \tag{12.6}$$

The qualitative behavior of the average Hamming distance can be derived from the above. At the beginning $\widehat{H}^{(t)}$ has a linear growth in time:

$$\widehat{H}^{(t)} \approx t. \tag{12.7}$$

For large t the expected Hamming distance $\widehat{H}^{(t)}$ tends to its asymptotic value of $n/2$ in an exponential way, with a "time constant" $\tau = n/2$

Let us now compare the evolution of the mean Hamming distance for different algorithms. The analysis is started as soon as the first local optimum is encountered by LS, when diversification becomes crucial. LS$^+$ has the same evolution as LS with

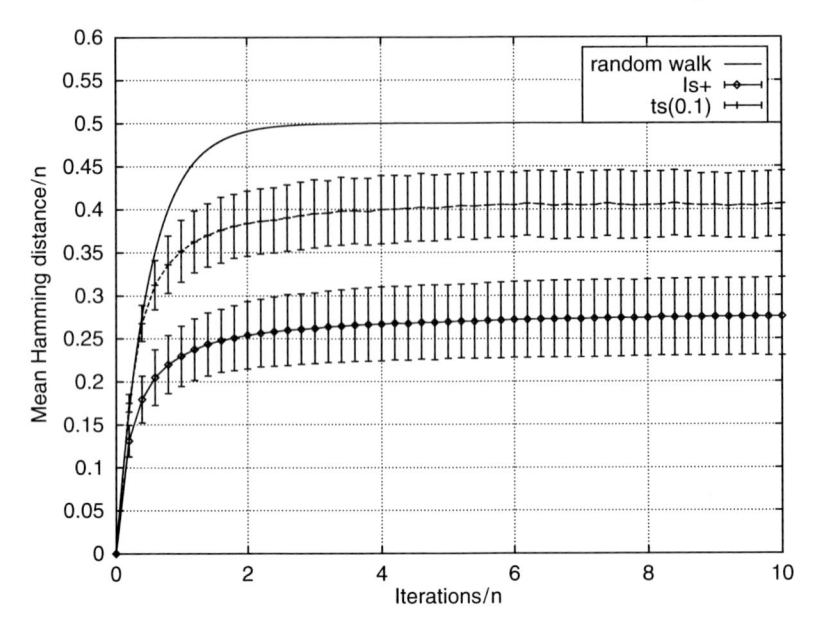

Fig. 12.3 Average Hamming distance reached by Random Walk, LS$^+$ and TS(0.1) from the first local optimum of LS, with standard deviation (MAX-3-SAT). Random walk evolution is also reported for reference

the only difference that it *always* moves to the best neighbor, even if the neighbor has a worse solution value f. LS$^+$, and Fixed-TS with fractional prohibition T_f equal to 0.1, denoted as TS(0.1), are then run for $10\,n$ additional iterations. Figure 12.3 shows the average Hamming distance as a function of the additional iterations after reaching the LS optimum, see [28] for experimental details.

Although the initial linear growth is similar to that of Random Walk, the Hamming distance does not reach the asymptotic value $n/2$ and a remarkable difference is present for the two algorithms. The fact that the asymptotic value is not reached even for large iteration numbers implies that all visited strings tend to lie in a confined region of the search space, with bounded Hamming distance from the starting point.

Let us note that, for large n values, most binary strings are at distance of approximately $n/2$ from a given string. In detail, the Hamming distances are distributed with a binomial distribution with the same probability of success and failure ($p = q = 1/2$): the fraction of strings at distance H is equal to

$$\binom{n}{H} \times \frac{1}{2^n}. \tag{12.8}$$

It is well known that the mean is $n/2$ and the standard deviation is $\sigma = \sqrt{n}/2$. The above coefficients increase up to the mean $n/2$ and then decrease. Because the ratio σ/n tends to zero for n tending to infinity, for large n values most strings are clustered in a narrow peak at Hamming distance $H = n/2$. As an example, one can use the Chernoff bound [119]:

12.4 Measuring Local Search Components: Diversification and Bias

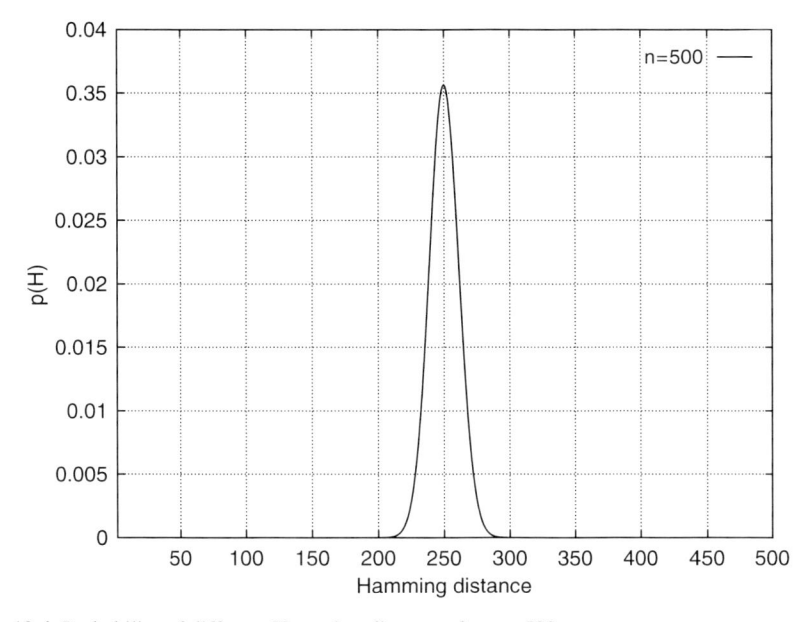

Fig. 12.4 Probability of different Hamming distances for $n = 500$

$$Pr[H \leq (1-\theta)pn] \leq e^{-\theta^2 np/2} \quad (12.9)$$

the probability to find a point at a distance less than $np = n/2$ decreases in the above exponential way ($\theta \geq 0$). The distribution of Hamming distances for $n = 500$ is shown in Fig. 12.4.

Clearly, if better local optima are located in a cluster that is not reached by the trajectory, they will never be found. In other words, a robust algorithm demands that some stronger diversification action is executed. For example, an option is to activate a restart after a number of iterations that is a small multiple of the time constant $n/2$.

12.4.1 The Diversification–Bias Compromise (D–B Plots)

When a local search component is started, new configurations are obtained at each iteration until the first local optimum is encountered, because the number of satisfied clauses increases by at least one. During this phase, additional diversification schemes are not necessary and potentially dangerous, because they could lead the trajectory astray, away from the local optimum.

The compromise between bias and diversification becomes critical after the first local optimum is encountered. In fact, if the local optimum is strict, the application of a move will worsen the f value, and an additional move could be selected to bring the trajectory back to the starting local optimum.

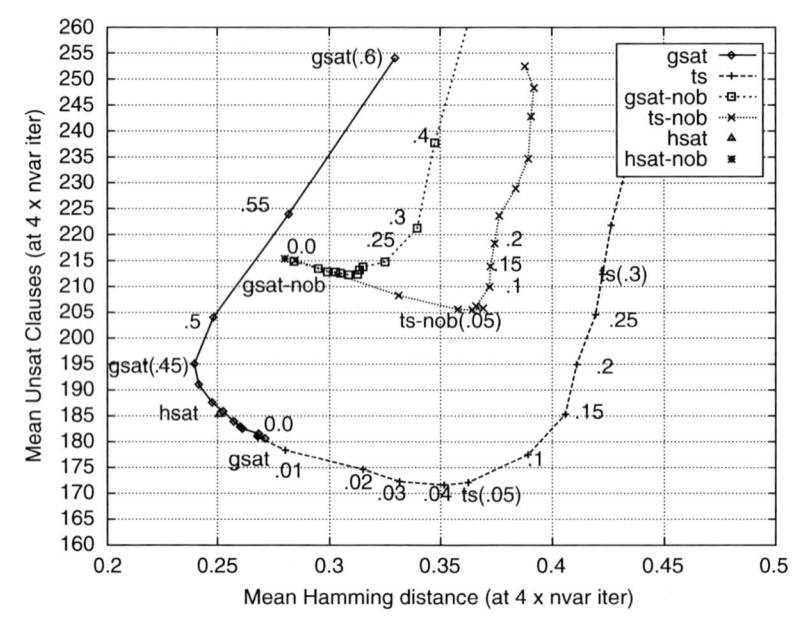

Fig. 12.5 Diversification–bias plane. Mean number of unsatisfied clauses after $4n$ iterations versus mean Hamming distance. MAX-3-SAT tasks. Each run starts from GSAT local optimum, see [28]

The mean bias and diversification depend on the value of the internal parameters of the different components. All runs proceed as follows: as soon as the first local optimum is encountered by LS, it is stored and the selected component is then run for additional $4n$ iterations. The final Hamming distance H from the stored local optimum and the final value of the number of unsatisfied clauses u are collected. The values are then averaged over different tasks and different random number seeds.

Different diversification–bias (D–B) plots are shown in Fig. 12.5. Each point gives the D–B coordinates $(\widehat{H_n}, \widehat{u})$, i.e., average Hamming distance divided by n and average number of unsatisfied clauses, for a specific parameter setting in the different algorithms. The Hamming distance is normalized with respect to the problem dimension n, i.e., $\widehat{H_n} \equiv \widehat{H}/n$. Three basic algorithms are considered: GSAT-with-walk, Fixed-TS, and HSAT. For each of these, two options about the guiding functions are studied: one adopts the "standard" oblivious function, the other the nonoblivious f_{NOB} introduced in Sect. 5.2.1. Finally, for GSAT-with-walk one can change the probability parameter p, while for Fixed-TS one can change the fractional prohibition T_f: parametric curves as a function of a single parameter are therefore obtained.

GSAT, Fixed-TS(0.0), and GSAT-with-walk(0.0) coincide: no prohibitions are present in TS and no stochastic choice is present in GSAT-with-walk. The point is marked with "0.0" in Fig. 12.5. By considering the parametric curve for GSAT-with-walk(p) (label "gsat" in Fig. 12.5), one observes a gradual increase of $\widehat{u}$ for increasing p, while the mean Hamming distance reached at first decreases and then increases. The initial decrease is unexpected because it contradicts the intuitive argument that more stochasticity implies more diversification. The reason for the above result is that there are two sources of "randomness" in the GSAT-with-walk algorithm (Fig. 5.2), one deriving from the random choice among variables in unsatisfied clauses, active with probability p, the other one deriving from the random breaking of ties if more variables achieve the largest Δf.

Because the first randomness source increases with p, the decrease in $\widehat{H_n}$ could be explained if the second source decreases. This conjecture has been tested and confirmed [28]. The larger amount of stochasticity implied by a larger p keeps the trajectory on a rough terrain at higher values of f, where flat portions tend to be rare. Vice versa, almost no tie is present when the nonoblivious function is used. The algorithm on the optimal frontier of Fig. 12.5 is Fixed-TS(T_f), and the effect of a simple *aspiration criterion* [107], and a *tie-breaking rule* for it is studied in [28].

The advantage of the D–B plot analysis is clear: it suggests possible causes for the behavior of different algorithms, leading to a more focused investigation.

12.4.2 A Conjecture: Better Algorithms are Pareto-Optimal in D–B Plots

A conjecture about the relevance of the diversification–bias metric is proposed in [28]. A relation of partial order, denoted by the symbol $\geq$ and called "domination," is introduced in a set of algorithms in the following way: Given two component algorithms A and B, A *dominates* B ($A \geq B$) if and only if it has a larger or equal diversification and bias: $\widehat{f_A} \geq \widehat{f_B}$ and $\widehat{H_{nA}} \geq \widehat{H_{nB}}$.

By definition, component A is a *maximal element* of the given relation if the other components in the set *do not* possess both a higher diversification, and a better bias. In the graph, one plots the number of *un*satisfied clauses vs. the Hamming distance, therefore the *maximal* components are in the lower-right corner of the set of $(\widehat{H_n}, \widehat{u})$ points. The points are characterized by the fact that no other point has both a larger diversification and a smaller number of satisfied clauses.

Conjecture
If local-search-based components are used in heuristic algorithms for optimization, the components producing the best f values during a run, on the average, are the maximal elements in the diversification–bias plane for the given partial order.

The conjecture produces some "falsifiable" predictions that can be tested experimentally. In particular, a partial ordering of the different components is introduced: component A is better than component B if $\widehat{H_{nA}} \geq \widehat{H_{nB}}$ and $\widehat{u_A} \leq \widehat{u_B}$. The ordering is partial because no conclusions can be reached if, for example, A has better diversification but worse bias when compared with B.

Clearly, when one applies a technique for optimization, one wants to maximize the best value found during the run. This value is affected by both the *bias* and the *diversification*. The search trajectory must visit preferentially points with large f values but, as soon as one of this point is visited, the search must proceed to visit new regions. The above conjecture is tested experimentally in [28], with fully satisfactory results. We do not expect this conjecture to be always valid but it is a useful guide when designing and understanding component algorithms.

A definition of three metrics is used in [224, 239] for studying algorithms for SAT and CSP. The first two metrics *depth* (average unsatisfied clauses) and *mobility* (Hamming distance speed) correspond closely to the above used *bias* and *diversification*. The third measure (*coverage*) takes a more global view at the search progress. In fact, one may have a large mobility but nonetheless remain confined in a small portion of the search space. A two-dimensional analogy is that of bird flying at high speed along a circular trajectory: if fresh corn is not on the trajectory, it will never discover it. Coverage is intended to measure *how systematically* the search explores the entire space. In other words, coverage is what one needs to ensure that eventually the optimal solution will be identified, no matter how it is camouflaged in the search landscape. The motivation for a *speed of coverage* measure is intuitively clear, but the detailed definition and implementation is somewhat challenging. In [224] a worst-case scenario is considered and *coverage* is defined as the size of the *largest unexplored gap in the search space*. For a binary string, this is given by the maximum Hamming distance between any unexplored assignment and the nearest explored assignment.

Unfortunately, measuring the speed of coverage it is not so fast, actually it can be NP-hard for problems with binary strings, and one has to resort to approximations [224]. For an example, one can consider sample points given by the negation of the visited points along the trajectory and determine the maximum and minimum distance between these points and points along the search trajectory. The rationale for this heuristic choice is that the negation of a string is the farthest point from a *given* string (one tries to be on the safe side to estimate the real coverage). After this estimate is available one divides by the number of search steps. Alternatively, one could consider how fast coverage decreases during the search (a discrete approximation of the coverage speed). Dual measures on the constraints are studied in [239].

Chapter 13
Open Problems

> *We are continually faced with a series of great opportunities brilliantly disguised as insoluble problems.*
> *(John W. Gardner)*

> *We've heard that a million monkeys at a million keyboards could produce this book; now, we know that is not true.*
> *(The Authors)*

We are at the conclusion of our journey and it is time to tell our seven readers about some future work which is waiting for brilliant and courageous pioneers. Needless to say, this is the less complete and less defined part of our book, we are dealing here with the open problem of defining the open problems. In the interest of brevity, we will be brutal, therefore do not take us at face value.

We are living in a time when most interesting basic optimization problems have at least one solution technique proposed. In fact, for most relevant problems, there are tens and tens of different proposals. Does it mean that there is no more serious and ground-breaking research work to be done? Does it mean that what is left are incremental adjustments leading to some percent improvement? Absolutely not.

In fact, when one considers how optimization is used in the real world, the picture is far from being all roses. *Simplifying life for the final user* and dealing with *optimization in the large* are the two key issues here. It is foreseeable that optimization will undergo a similar process to that occurred to other computer science and engineering areas: from tools requiring expert users and costly deployment times, to more and more automated tools asking less and less technical questions to the final user so that he avoids a cognitive overload and concentrates his intellectual power on the crucial issues for his job. After all, no Ph.D. is required to browse the Web and obtain relevant information, no Ph.D. should be required to optimize the processes and operations of your business.

The foreseeable increase in the automation and the development of simpler and focussed user interfaces will not come for free. After all, there are no simple solutions to complex problems, and the complexity which will hopefully be reduced for the final user will be *shifted* to the algorithms. Major and ground-breaking research efforts are required in this area; the road toward the dream of self-programming computers is a long and difficult one, and the goal can elude also the future generations. Both theoretical and engineering work is ahead requesting future teams of solid researchers with a background in different areas, ranging from algorithms, to software engineering, operations research, machine learning and artificial intelligence, statistics, human–computer interfaces.

Finally, let us present a list of some crucial issues to be solved.

- *Optimization in the large:* The term is inspired by *programming in the large* in software development, which involves programming by large groups of people or by small groups over long time periods. To avoid complicated programs that can be challenging for maintainers to understand, emphasis on modular design and organization with precisely-specified interactions will be crucial. This *divide et impera* approach is particularly critical for intelligent optimization and reactive search heuristics, where different components are combined, integrated, modified by off-line or on-line learning schemes. Algorithm selection, adaptation, integration, development of comprehensive solutions are on the agenda.
- *Simplify life for the final user!* Presenting to the final user the relevant information for effective decision making is crucial, as well as reacting in a rapid manner to the real user needs, either explicit or implicit in his actions. Human–computer interaction aspects are important, but also ways to adapt the problem solving process in time, as soon as more information is derived from the user.
- *Optimize the problem definition!* It is a rule of thumb that 90% of the time is spent on defining the problem, on defining the f function to be optimized, its solution requiring in many cases a negligible effort. By the way, maybe this explains why many apparently suboptimal optimization schemes are still very popular and publicized by smart marketing strategies. If the real problem is the modeling part, more theoretical advancements and tools should be dedicated to it, for example to help the user to identify the real objectives and constraints, and the real tradeoffs to be considered when multiple objectives are present, which is a very frequent case in the real world.
- *Diversity of solutions:* It is now possible to generate many alternative solutions to a problem, in same cases tons of them. This opens opportunities for a better and more informed choice, provided that the information bottleneck at the user is taken into account. Providing a list of thousands of solution vectors is clearly unacceptable. Proper data-mining methods, clustering techniques, and graphical presentations have to be used, so that the user can focus on the relevant issues and make an informed decision.
- *Interactivity and user adaptation:* A manager wants and need to be in charge of the final choice, in a world where constraints and priorities change in a rapid manner. The algorithms and software tools must follow the user changes through interactivity, learning his changing preferences, and rapidly adapting to his needs.
- *Unification of intelligent optimization schemes:* Good science proceeds by providing an unified explanation of seemingly different phenomena. In fact, scientific explanation in some cases coincides with unification. The picture is not very positive in the wide area of intelligent optimization: many superficially different strategies are in some cases deeply related. A stronger effort is required to focus more on the phenomena and less on the superficial differences and different terminologies, in some cases related to marketing issues more than scientific ones. An example is given by the many related ways to "escape from local minima" described in the book.

13 Open Problems

- *Exploration vs. exploitation balance:* Aka intensification/diversification dilemma. It is one of the most critical and less understood problems. Reinforcement learning has been used to aim at an automated balance, but this does not solve the problem in a satisfactory manner. In fact, the same plague also affects RL, where again early intensification risks not identifying a better but unexplored solution. Theoretical ways of dealing with the dilemma, in addition to the simplified k-armed bandit toy examples, are badly need.
- *Policies vs. implementations:* A distinction between the *what* and the *how* is always helpful for understanding and designing. As an example, focus on local search trajectories if you want to relate the amount of information derived from the fitness surface to the optimization success, focus on the supporting data structures if you are interested in the final CPU times. In the experimental comparisons, both aspects are important but they need to be clearly separated.
- *Metrics, metrics, metrics:* There is no science without measurements, but defining relevant quantities to measure which have predictive power for problem complexity, runtime, etc. is not simple. Automated machine learning techniques can help, including methods to select features, but metrics defined by the researcher are in most cases more directly related to a human-understandable explanation. On the contrary, parameter-free models in machine learning have practical relevance but encounter difficulties in explaining why they work. A combination of high-level (symbolic) explanation with some low-level (subsymbolic) modules can be a way to get the best of both worlds.
- *Experimental computer science:* See also above. When asked – Do you have a mathematical proof for the superiority of this heuristic? – just answer – Do you have a mathematical proof that the law of gravity holds? Computer science, and intelligent optimization is no exception, is becoming an experimental science like it previously happened to Physics, with the additional excitement that the real-world in physics has already been defined, while the researcher can play god in this framework and define new worlds to be analyzed. Of course, mathematics can help, see statistics, not in the traditional way of proving theorems about algorithms but as a tool in the inductive process. Design of experiment is an important issue, it is not sufficient to bombard the user with lots of data. Asking the right questions, explaining the experimental process, analyzing with statistically sound methods, and providing honest conclusions are the bread and butter of experimental researchers.

This is the end, my friend, but the story continues online at the Web sites cited in the preface. Or contact the authors directly if you have relevant optimization problems waiting for effective solutions.

References

1. Aarts, E., Korst, J.: Boltzmann machines for travelling salesman problems. European Journal of Operational Research **39**, 79–95 (1989)
2. Aarts, E.H.L., Korst, J., Zwietering, P.: Deterministic and randomized local search. In: M.C. Mozer, P. Smolensky, D.E. Rumelhart (eds.) Mathematical Perspectives on Neural Networks. Lawrence Erlbaum, Hillsdale, NJ (1995)
3. Abramson, D., Dang, H., Krisnamoorthy, M.: Simulated annealing cooling schedules for the school timetabling problem. Asia-Pacific Journal of Operational Research **16**, 1–22 (1999). URL citeseer.ist.psu.edu/article/abramson97simulated.html
4. Aho, A.V., Hopcroft, J.E., Ullman, J.D.: Data Structures and Algorithms. Addison-Wesley, Reading, MA (1983)
5. Alimonti, P.: New local search approximation techniques for maximum generalized satisfiability problems. In: Proceedings of the Second Italian Conference on Algorithms and Complexity, pp. 40–53 (1994)
6. Altenberg, L.: Fitness distance correlation analysis: An instructive counterexample. Proceedings of the 7th International Conference on Genetic Algorithms (ICGA97), pp. 57–64 (1997)
7. Anderson, B., Moore, A., Cohn, D.: A nonparametric approach to noisy and costly optimization. In: Proceedings of the Seventeenth International Conference on Machine Learning (ICML 2000), pp. 17–24 (2000)
8. Anzellotti, G., Battiti, R., Lazzizzera, I., Soncini, G., Zorat, A., Sartori, A., Tecchiolli, G., Lee, P.: Totem: A highly parallel chip for triggering applications with inductive learning based on the reactive tabu search. International Journal of Modern Physics C **6**(4), 555–560 (1995)
9. Atkeson, C.G., Schaal, S.A., Moore, A.: Locally weighted learning. AI Review **11**, 11–73 (1997)
10. Auer, P., Cesa-Bianchi, N., Fischer, P.: Finite-time analysis of the multiarmed bandit problem. Machine Learning **47**(2/3), 235–256 (2002). URL citeseer.ist.psu.edu/auer00finitetime.html
11. Bahren, J., Protopopescu, V., Reister, D.: Trust: A deterministic algorithm for global optimization. Science **276**, 10941,097 (1997)
12. Baluja, S.: Using a priori knowledge to create probabilistic models for optimization. International Journal of Approximate Reasoning **31**(3), 193–220 (2002)
13. Baluja, S., Barto, A., Boyan, K.B.J., Buntine, W., Carson, T., Caruana, R., Cook, D., Davies, S., Dean, T., et al.: Statistical machine learning for large-scale optimization. Neural Computing Surveys **3**, 1–58 (2000)
14. Baluja, S., Caruana, R.: Removing the genetics from the standard genetic algorithm. Tech. Rep. CMU-CS-95-141, School of Computer Science, Carnegie Mellon University, Pittsburgh, PA (1995)

15. Barr, R.S., Golden, B.L., Kelly, J.P., Resende, M.G.C., Stewart, W.: Designing and reporting on computational experiments with heuristic methods. Journal of Heuristics **1**(1), 9–32 (1995)
16. Barricelli, N.: Numerical testing of evolution theories. Acta Biotheoretica **16**(1), 69–98 (1962)
17. Barto, A.G., Sutton, R.S., Anderson, C.W.: Neurolike adaptive elements that can solve difficult learning problems. IEEE Transactions on Systems, Man and Cybernetics **13**, 834–846 (1983)
18. Battiti, R.: First-and second-order methods for learning: Between steepest descent and newton's method. Neural Computation **4**, 141–166 (1992)
19. Battiti, R.: Using the mutual information for selecting features in supervised neural net learning. IEEE Transactions on Neural Networks **5**(4), 537–550 (1994)
20. Battiti, R.: Partially persistent dynamic sets for history-sensitive heuristics. Tech. Rep. UTM-96-478, Dip. di Matematica, Universita' di Trento, Trento, Italy (1996). Revised version, Presented at the Fifth DIMACS Challenge, Rutgers, NJ, 1996
21. Battiti, R.: Reactive search: Toward self-tuning heuristics. In: V.J. Rayward-Smith, I.H. Osman, C.R. Reeves, G.D. Smith (eds.) Modern Heuristic Search Methods, chap. 4, pp. 61–83. Wiley, New York, NY (1996)
22. Battiti, R.: Time- and space-efficient data structures for history-based heuristics. Tech. Rep. UTM-96-478, Dip. di Matematica, Universita' di Trento, Trento, Italy (1996)
23. Battiti, R., Bertossi, A.A.: Greedy, prohibition, and reactive heuristics for graph partitioning. IEEE Transactions on Computers **48**(4), 361–385 (1999)
24. Battiti, R., Brunato, M.: Reactive search for traffic grooming in WDM networks. In: S. Palazzo (ed.) Evolutionary Trends of Internet, IWDC2001, Taormina, LNCS, vol. 2170, pp. 56–66. Springer, Berlin (2001)
25. Battiti, R., Colla, A.M.: Democracy in neural nets: Voting schemes for accuracy. Neural Networks **7**(4), 691–707 (1994)
26. Battiti, R., Tecchiolli, G.: Parallel biased search for combinatorial optimization: Genetic algorithms and tabu. Microprocessor and Microsystems **16**, 351–367 (1992)
27. Battiti, R., Protasi, M.: Solving max-sat with non-oblivious functions and history-based heuristics. Tech. Rep., Dipartimento di Matematica, Unversita' di Trento, Trento, Italy (1996)
28. Battiti, R., Protasi, M.: Reactive search, a history-sensitive heuristic for MAX-SAT. ACM Journal of Experimental Algorithmics **2**(ARTICLE 2) (1997). http://www.jea.acm.org/
29. Battiti, R., Protasi, M.: Solving MAX-SAT with non-oblivious functions and history-based heuristics. In: D. Du, J. Gu, P.M. Pardalos (eds.) Satisfiability Problem: Theory and Applications, no. 35 in DIMACS: Series in Discrete Mathematics and Theoretical Computer Science, pp. 649–667. American Mathematical Society, Association for Computing Machinery, New York, NY (1997)
30. Battiti, R., Protasi, M.: Reactive local search techniques for the maximum k-conjunctive constraint satisfaction problem. Discrete Applied Mathematics **96–97**, 3–27 (1999)
31. Battiti, R., Protasi, M.: Reactive local search for the maximum clique problem. Algorithmica **29**(4), 610–637 (2001)
32. Battiti, R., Tecchiolli, G.: Learning with first, second, and no derivatives: A case study in high energy physics. Neurocomputing **6**, 181–206 (1994)
33. Battiti, R., Tecchiolli, G.: The reactive tabu search. ORSA Journal on Computing **6**(2), 126–140 (1994)
34. Battiti, R., Tecchiolli, G.: Simulated annealing and tabu search in the long run: A comparison on QAP tasks. Computer and Mathematics with Applications **28**(6), 1–8 (1994)
35. Battiti, R., Tecchiolli, G.: Local search with memory: Benchmarking rts. Operations Research Spektrum **17**(2/3), 67–86 (1995)
36. Battiti, R., Tecchiolli, G.: Training neural nets with the reactive tabu search. IEEE Transactions on Neural Networks **6**(5), 1185–1200 (1995)
37. Battiti, R., Tecchiolli, G.: The continuous reactive tabu search: Blending combinatorial optimization and stochastic search for global optimization. Annals of Operations Research – Metaheuristics in Combinatorial Optimization **63**, 153–188 (1996)

References

38. Baxter, J.: Local optima avoidance in depot location. The Journal of the Operational Research Society **32**(9), 815–819 (1981)
39. Bayer, R.: Symmetric binary b-trees: Data structure and maintenance algorithms. Acta Informatica **1**, 290–306 (1972)
40. Bennett, K., Parrado-Hernández, E.: The interplay of optimization and machine learning research. The Journal of Machine Learning Research **7**, 1265–1281 (2006)
41. Bentley, J.: Experiments on traveling salesman heuristics. In: Proceedings of the first annual ACM-SIAM symposium on Discrete algorithms, pp. 91–99 (1990)
42. Bertsekas, D., Tsitsiklis, J.: Neuro-Dynamic Programming. Athena Scientific, Belmont, MA (1996)
43. Biazzini, M., Brunato, M., Montresor, A.: Towards a decentralized architecture for optimization. In: Proceedings of the Twenty-Second IEEE International Parallel and Distributed Processing Symposium. Miami, FL (2008)
44. Birattari, M., Stützle, T., Paquete, L., Varrentrapp, K.: A racing algorithm for configuring metaheuristics. In: W. Langdon et al. (ed.) Proceedings of the Genetic and Evolutionary Computation Conference, pp. 11–18. Morgan Kaufmann, San Francisco, CA (2002). Also available as: AIDA-2002-01 Technical Report of Intellektik, Technische Universität Darmstadt, Darmstadt, Germany
45. Blum, C., Roli, A.: Metaheuristics in combinatorial optimization: Overview and conceptual comparison. ACM Computing Surveys **35**(3), 268–308 (2003). DOI http://doi.acm.org/10.1145/937503.937505
46. de Boer, P., Kroese, D., Mannor, S., Rubinstein, R.: A tutorial on the cross-entropy method. Annals of Operations Research **134**, 19–67 (2005). URL citeseer.ist.psu.edu/deboer02tutorial.html
47. Boese, K., Kahng, A., Muddu, S.: On the big valley and adaptive multi-start for discrete global optimizations. Operation Research Letters **16**(2) (1994)
48. Bollobás: Random Graphs. Cambridge University Press, Cambridge (2001)
49. de Bonet, J.S., Jr., Isbell, C.L., Viola, P.: MIMIC: Finding optima by estimating probability densities. In: M.C. Mozer, M.I. Jordan, T. Petsche (eds.) Advances in Neural Information Processing Systems, vol. 9, p. 424. The MIT Press, Cambridge, MA (1997). URL citeseer.ist.psu.edu/debonet96mimic.html
50. Boyan, J., Moore, A.: Using prediction to improve combinatorial optimization search. In: Proceedings of AISTATS-6 (1997)
51. Boyan, J., Moore, A.: Learning evaluation functions to improve optimization by local search. The Journal of Machine Learning Research **1**, 77–112 (2001)
52. Boyan, J.A., Moore, A.W.: Learning evaluation functions for global optimization and boolean satisfability. In: A. Press (ed.): Proceedings of the Fifteenth National Conference on Artificial Intelligence (AAAI), pp. 3–10 (1998)
53. Braysy, O.: A reactive variable neighborhood search for the vehicle-routing problem with time windows. INFORMS Journal on Computing **15**(4), 347–368 (2003)
54. Breimer, E., Goldberg, M., Hollinger, D., Lim, D.: Discovering optimization algorithms through automated learning. In: Graphs and Discovery, DIMACS Series in Discrete Mathematics and Theoretical Computer Science, vol. 69, pp. 7–27. American Mathematical Society (2005)
55. Brent, R.P.: Algorithms for Minimization Without Derivatives. Prentice-Hall, Englewood Cliffs, NJ (1973)
56. Brunato, M., Battiti, R.: Statistical learning theory for location fingerprinting in wireless LANs. Computer Networks **47**(6), 825–845 (2005)
57. Brunato, M., Battiti, R.: The reactive affine shaker: A building block for minimizing functions of continuous variables. Tech. Rep. DIT-06-012, Università di Trento, Trento, Italy (2006)
58. Brunato, M., Battiti, R., Pasupuleti, S.: A memory-based rash optimizer. In: A.F.R.H.H. Geffner (ed.) Proceedings of AAAI-06 Workshop on Heuristic Search, Memory Based Heuristics and Their Applications, pp. 45–51. Boston, MA (2006). ISBN 978-1-57735-290-7

59. Burges, C.J.C.: A tutorial on support vector machines for pattern recognition. Data Mining and Knowledge Discovery **2**(2), 121–167 (1998)
60. Carchrae, T., Beck, J.: Applying machine learning to low-knowledge control of optimization algorithms. Computational Intelligence **21**(4), 372–387 (2005)
61. Cheeseman, P., Kanefsky, B., Taylor, W.: Where the really hard problems are. In: Proceedings of the Twelfth IJCAI, pp. 331–337 (1991)
62. Chelouah, R., Siarry, P.: Tabu search applied to global optimization. European Journal of Operational Research **123**, 256–270 (2000)
63. Cherny, V.: A thermodynamical approach to the traveling salesman problem: An efficient simulation algorithm. Journal of Optimization Theory and Applications **45**, 41–45 (1985)
64. Chiang, T.S., Chow, Y.: On the convergence rate of annealing processes. SIAM Journal on Control and Optimization **26**(6), 1455–1470 (1988)
65. Chiang, W., Russell, R.: A reactive tabu search metaheuristic for the vehicle routing problem with time windows. INFORMS Journal on Computing **9**, 417–430 (1997)
66. Cicirello, V.: Boosting Stochastic Problem Solvers Through Online Self-Analysis of Performance. Ph.D. thesis, Carnegie Mellon University, Pittsburgh, PA (2003) Also available as Technical Report CMU-RI-TR-03-27
67. Cicirello, V., Smith, S.: The max k-armed bandit: A new model for exploration applied to search heuristic selection. In: Twentieth National Conference on Artificial Intelligence (AAAI-05) (2005). Best Paper Award
68. Cicirello, V.A., Smith, S.F.: Principles and practice of constraint programming CP 2004, LNCS, vol. 3258, chap. Heuristic Selection for Stochastic Search Optimization: Modeling Solution Quality by Extreme Value Theory, pp. 197–211. Springer, Berlin/Heidelberg (2004)
69. Clark, D.A., Frank, J., Gent, I.P., MacIntyre, E., Tomov, N., Walsh, T.: Local search and the number of solutions. In: Principles and Practice of Constraint Programming, pp. 119–133 (1996). URL citeseer.ist.psu.edu/article/clark96local.html
70. Clerc, M., Kennedy, J.: The particle swarm – Explosion, stability, and convergence in a multidimensional complex space. IEEE Transactions on Evolutionary Computation **6**(1), 58–73 (2002)
71. Cleveland, W.S., Devlin, S.J.: Locally-weighted regression: An approach to regression analysis by local fitting. Journal of the American Statistical Association **83**, 596–610 (1988)
72. Codenotti, B., Manzini, G., Margara, L., Resta, G.: Perturbation: An efficient technique for the solution of very large instances of the euclidean tsp. INFORMS Journal on Computing **8**(2), 125–133 (1996)
73. Connolly, D.: An improved annealing scheme for the QAP. European Journal of Operational Research **46**(1), 93–100 (1990)
74. Conover, W.J.: Practical Nonparametric Statistics. Wiley, New York, NY (1999). URL http://www.amazon.co.uk/exec/obidos/ASIN/0471160687/citeulike-21
75. Corana, A., Marchesi, M., Martini, C., Ridella, S.: Minimizing multimodal functions of continuous variables with the simulated annealing algorithm. ACM Transactions on Mathematical Software **13**(3), 262–280 (1987). DOI http://doi.acm.org/10.1145/29380.29864
76. Cormen, T.H., Leiserson, C.E., Rivest, R.L.: Introduction to Algorithms. McGraw-Hill, New York (1990)
77. Cox, B.J.: Object Oriented Programming, an Evolutionary Approach. Addison-Wesley, Reading, MA (1990)
78. Crainic, T., Toulouse, M.: Parallel strategies for metaheuristics. In: F. Glover, G. Kochenberger (eds.) State-of-the-Art Handbook in Metaheuristics, chap. 1. Kluwer, Dordrecht (2002)
79. Crawford, J.M., Auton, L.D.: Experimental results on the crossover point in random 3-sat. Artificial Intelligence **81**(1–2), 31–57 (1996). DOI http://dx.doi.org/10.1016/0004-3702(95)00046-1
80. Daida, J., Yalcin, S., Litvak, P., Eickhoff, G., Polito, J.: Of metaphors and darwinism: Deconstructing genetic programmings chimera. In: Proceedings CEC-99: Congress in Evolutionary Computation, Piscataway, p. 453462. IEEE Press, Piscataway, NY (1999)

81. Dammeyer, F., Voss, S.: Dynamic tabu list management using the reverse elimination method. Annals of Operations Research **41**, 31–46 (1993)
82. Darwin, C.: On The Origin of Species. Signet Classic, reprinted 2003 (1859)
83. Dawkins, R.: The Selfish Gene. Oxford University, Oxford (1976)
84. Delmaire, H., Dıaz, J., Fernandez, E., Ortega, M.: Reactive GRASP and Tabu Search based heuristics for the single source capacitated plant location problem. INFOR **37**, 194–225 (1999)
85. Demers, A., Greene, D., Hauser, C., Irish, W., Larson, J., Shenker, S., Sturgis, H., Swinehart, D., Terry, D.: Epidemic algorithms for replicated database maintenance. In: Proceedings of the Sixth Annual ACM Symposium on Principles of Distributed Computing (PODC'87), pp. 1–12. ACM Press, Vancouver, British Columbia, Canada (1987). URL http://doi.acm.org/10.1145/41840.41841
86. Dorigo, M., Blum, C.: Ant colony optimization theory: A survey. Theoretical Computer Science **344**(2–3), 243–278 (2005). DOI http://dx.doi.org/10.1016/j.tcs.2005.05.020
87. Driscoll, J.R., Sarnak, N., Sleator, D.D., Tarjan, R.E.: Making data structures persistent. In: Proceedings of the Eighteenth Annual ACM Symposium on Theory of Computing. ACM, Berkeley, CA (1986)
88. Dubrawski, A., Schneider, J.: Memory based stochastic optimization for validation and tuning of function approximators. In: Conference on AI and Statistics (1997)
89. Duda, R., Hart, P., Stork, D.: Pattern Classification. Wiley-Interscience, New York, NY (2000)
90. Eiben, A., Hinterding, R., Michalewicz, Z.: Parameter control in evolutionary algorithms. IEEE Transactions on Evolutionary Computation **3**(2), 124–141 (1999)
91. Eiben, A., Horvath, M., Kowalczyk, W., Schut, M.: Reinforcement learning for online control of evolutionary algorithms. In: S. Brueckner, M. Hassas, D. Yamins Jelasity, (eds.) Proceedings of the Fourth International Workshop on Engineering Self-Organizing Applications (ESOA'06), LNAI. Springer, Berlin (2006)
92. Erdos, P., Renyi, A.: On random graphs. Publicationes Mathematicae Debrecen **6**, 290–297 (1959)
93. Faigle, U., Kern, W.: Some convergence results for probabilistic tabu search. ORSA Journal on Computing **4**(1), 32–37 (1992)
94. Feo, T., Resende, M.: Greedy randomized adaptive search procedures. Journal of Global Optimization **6**, 109–133 (1995)
95. Ferreira, A., Zerovnik, J.: Bounding the probability of success of stochastic methods for global optimization. Computers Mathematics with Applications **25**(10/11), 1–8 (1993)
96. Fleischer, M.A.: Cybernetic optimization by simulated annealing: Accelerating convergence by parallel processing and probabilistic feedback control. Journal of Heuristics **1**(2), 225–246 (1996)
97. Fong, P.W.L.: A quantitative study of hypothesis selection. In: International Conference on Machine Learning, pp. 226–234 (1995). URL citeseer.ist.psu.edu/fong95quantitative.html
98. Fortz, B., Thorup, M.: Increasing internet capacity using local search. Computational Optimization and Applications **29**(1), 13–48 (2004)
99. Frank, J.: Weighting for godot: Learning heuristics for GSAT. In: Proceedings of the National Conference on Artificial Intelligence, vol. 13, pp. 338–343. Wiley, New York, NY (1996)
100. Frank, J.: Learning short-term weights for GSAT. In: Proceedings of the International Joint Conference on Artificial Intelligence, vol. 15, pp. 384–391. Lawrence Erlbaum Associates, Hillsdale, NJ (1997)
101. Fraser, A., Burnell, D.: Computer models in genetics. McGraw-Hill, New York (1970)
102. Fraser, A.M., Swinney, H.L.: Independent coordinates for strange attractors from mutual information. Physical Review A **33**(2), 1134–1140 (1986). DOI 10.1103/PhysRevA.33.1134
103. Gagliolo, M., Schmidhuber, J.: A neural network model for inter-problem adaptive online time allocation. In: W. Duch et al. (ed.) Proceedings Artificial Neural Networks: Formal Models and Their Applications – ICANN 2005, Fifteenth International Conference, vol. 2, pp. 7–12. Springer, Berlin, Warsaw (2005)

104. Gagliolo, M., Schmidhuber, J.: Dynamic algorithm portfolios. In: Proceedings AI and MATH'06, Ninth International Symposium on Artificial Intelligence and Mathematics. Fort Lauderdale, FL (2006)
105. Gagliolo, M., Schmidhuber, J.: Impact of censored sampling on the performance of restart strategies. In: CP 2006 – Twelfth International Conference on Principles and Practice of Constraint Programming – Nantes, France, pp. 167–181. Springer, Berlin (2006)
106. Gent, I., Walsh, T.: Towards an understanding of hill-climbing procedures for sat. In: Proceedings of the Eleventh National Conference on Artificial Intelligence, pp. 28–33. AAAI Press/The MIT Press, San Jose, CA/Cambridge, MA (1993)
107. Glover, F.: Tabu search – part i. ORSA Journal on Computing **1**(3), 190–260 (1989)
108. Glover, F.: Tabu search – part ii. ORSA Journal on Computing **2**(1), 4–32 (1990)
109. Glover, F.: Tabu search: Improved solution alternatives. In: J.R. Birge, K.G. Murty (eds.) Mathematical Programming, State of the Art, pp. 64–92. University of Michigan, Press, Ann Arbor, MI (1994)
110. Glover, F.: Scatter search and star-paths: Beyond the genetic metaphor. Operations Research Spektrum **17**(2/3), 125–138 (1995)
111. Glover, F., Laguna, M., Marti, R.: Fundamentals of scatter search and path relinking. Control and Cybernetics **39**(3), 653–684 (2000)
112. Gomes, C., Selman, B., Crato, N., Kautz, H.: Heavy-tailed phenomena in satisfiability and constraint satisfaction problems. Journal of Automated Reasoning **24**(1/2), 67–100 (2000)
113. Gomes, C.P., Selman, B.: Algorithm portfolios. Artificial Intelligence **126**(1–2), 43–62 (2001). DOI http://dx.doi.org/10.1016/S0004-3702(00)00081-3
114. Gomes, F.C., Pardalos, P., Oliveira, C.S., Resende, M.G.C.: Reactive grasp with path relinking for channel assignment in mobile phone networks. In: DIALM'01: Proceedings of the Fifth international workshop on Discrete algorithms and methods for mobile computing and communications, pp. 60–67. ACM Press, New York, NY (2001). DOI http://doi.acm.org/10.1145/381448.381456
115. Greistorfer, P., VoS, S.: Controlled Pool Maintenance in Combinatorial Optimization, Operations Research/Computer Science Interfaces, vol. 30. Springer, Berlin (2005)
116. Gu, J.: Parallel algorithms and architectures for very fast ai search. Ph.D. thesis, University of Utah, Salt Lake City, UT (1989)
117. Gu, J.: Efficient local search for very large-scale satisfiability problem. ACM SIGART Bulletin **3**(1), 8–12 (1992)
118. Guibas, L.J., Sedgewick, R.: A dichromatic framework for balanced trees. In: Proceedings of the Nineteenth Annual Symposium on Foundations of Computer Science, pp. 8–21. IEEE Computer Society, Silver Spring, MD (1978)
119. Hagerup, T., Rueb, C.: A guided tour of chernoff bounds. Information Processing Letters **33**, 305–308 (1989/90)
120. Hajek, B.: Cooling schedules for optimal annealing. Mathematics of Operations Research **13**(2), 311–329 (1988)
121. Hamza, K., Mahmoud, H., Saitou, K.: Design optimization of N-shaped roof trusses using reactive taboo search. Applied Soft Computing Journal **3**(3), 221–235 (2003)
122. Hamza, K., Saitou, K., Nassef, A.: Design optimization of a vehicle b-pillar subjected to roof crush using mixed reactive taboo search. In: Proceedings of the ASME 2003 Design Engineering and Technical Conference, Chicago, IL, pp. 1–9 (2003)
123. Hansen, N.M.P.: Variable neighborhood search. Computers and Operations Research **24**(11), 1097–1100 (1997)
124. Hansen, P., Jaumard, B.: Algorithms for the maximum satisfiability problem. Computing **44**, 279–303 (1990)
125. Hansen, P., Mladenovic, N.: Variable neighborhood search. In: E. Burke, G. Kendall (eds.) Search methodologies: Introductory tutorials in optimization and decision support techniques, pp. 211–238. Springer, Berlin (2005)
126. Heckendorn, R.B., Rana, S., Whitey, D.L.: Test function generators as embedded landscapes. In: W. Banzhaf, C. Reeves (eds.) Foundations of Genetic Algorithms 5, pp. 183–198. Morgan Kaufmann, San Francisco (1999)

127. Hertz, J., Krogh, A., Palmer, R.: Introduction to the Theory of Neural Computation. Addison-Wesley, Redwood City, CA (1991). URL citeseer.ist.psu.edu/hertz91introduction.html
128. Hifi, M., Michrafy, M.: A reactive local search-based algorithm for the disjunctively constrained knapsack problem. Journal of the Operational Research Society **57**(6), 718–726 (2006)
129. Hifi, M., Michrafy, M., Sbihi, A.: A Reactive local search-based algorithm for the multiple-choice multi-dimensional knapsack problem. Computational Optimization and Applications **33**(2), 271–285 (2006)
130. Hinterding, R., Michalewicz, Z., Eiben, A.: Adaptation in evolutionary computation: A survey. In: IEEE International Conference on Evolutionary Computation, pp. 65–69 (1997)
131. Hinton, G., Nowlan, S.: How learning can guide evolution. Complex Systems **1**(1), 495–502 (1987)
132. Hogg, T.: Applications of Statistical Mechanics to Combinatorial Search Problems, vol. 2, pp. 357–406. World Scientific, Singapore (1995)
133. Hogg, T., Huberman, B.A., Williams, C.P.: Phase transitions and the search problem. Artificial Intelligence **81**(1-2), 1–15 (1996). DOI http://dx.doi.org/10.1016/0004-3702(95)00044-5
134. Holland, J.: Adaptation in Nature and Artificial Systems. University of Michigan Press, Ann Arbor, MI (1975)
135. Hooker, J.: Testing heuristics: We have it all wrong. Journal of Heuristics **1**(1), 33–42 (1995)
136. Hoos, H.: An adaptive noise mechanism for WalkSAT. In: Proceedings of the National Conference on Artificial Intelligence, vol. 18, pp. 655–660. AAAI Press MIT Press, San Jose, CA/Cambridge, MA (1999)
137. Hornik, K., Stinchcombe, M., White, H.: Multilayer feedforward networks are universal approximators. Neural Networks **2**(5), 359–366 (1989)
138. Horvitz, E., Ruan, Y., Gomes, C., Kautz, H., Selman, B., Chickering, D.M.: A bayesian approach to tackling hard computational problems. In: Seventeenth Conference on Uncertainty in Artificial Intelligence, Seattle, WA, pp. 235–244 (2001)
139. Hu, B., Raidl, G.R.: Variable neighborhood descent with self-adaptive neighborhood-ordering. In: C. Cotta, A.J. Fernandez, J.E. Gallardo (eds.) Proceedings of the Seventh EU/MEeting on Adaptive, Self-Adaptive, and Multi-Level Metaheuristics, Malaga, Spain (2006)
140. Huang, M., Romeo, F., Sangiovanni-Vincentelli, A.: An efficient general cooling schedule for simulated annealing. In: IEEE International Conference on Computer Aided Design, pp. 381–384 (1986)
141. Huberman, B., Hogg, T.: Phase transitions in artificial intelligence systems. Artificial Intelligence **33**(2), 155–171 (1987)
142. Huberman, B.A., Lukose, R.M., Hogg, T.: An economics approach to hard computational problems. Science **275**, 51–54 (1997)
143. Hutter, F., Hamadi, Y.: Parameter adjustment based on performance prediction: Towards an instance-aware problem solver. Tech. Rep. MSR-TR-2005-125, Microsoft Research, Cambridge, UK (2005)
144. Ingber, L.: Very fast simulated re-annealing. Mathematical and Computer Modelling **12**(8), 967–973 (1989)
145. Ishtaiwi, A., Thornton, J.R., A. Anbulagan, S., Pham, D.N.: Adaptive clause weight redistribution. In: Proceedings of the Twelfth International Conference on the Principles and Practice of Constraint Programming, CP-2006, Nantes, France, pp. 229–243 (2006)
146. Gu, J., Du, B.: A multispace search algorithm (invited paper). DIMACS Monograph on Global Minimization of Nonconvex Energy Functions (1996)
147. Jacquet, W., Truyen, B., de Groen, P., Lemahieu, I., Cornelis, J.: Global optimization in inverse problems: A comparison of Kriging and radial basis functions (2005). http://arxiv.org/abs/math/0506440
148. Janis, I.: Victims of groupthink. Houghton Mifflin (1972)

149. Joachims, T.: Making large-scale SVM learning practical. In: B. Schölkopf, C.J.C. Burges, A.J. Smola (eds.) Advances in Kernel Methods – Support Vector Learning, chap. 11. MIT Press, Cambridge, MA (1999)
150. Johnson, D.: Local optimization and the travelling salesman problem. In: Proceedings of the Seventeenth Colloquium on Automata Languages and Programming, LNCS, vol. 447. Springer, Berlin (1990)
151. Johnson, D.S., Aragon, C.R., McGeoch, L.A., Schevon, C.: Optimization by simulated annealing: An experimental evaluation. part i, graph partitioning. Oper. Res. **37**(6), 865–892 (1989)
152. Johnson, D.S., Aragon, C.R., McGeoch, L.A., Schevon, C.: Optimization by simulated annealing: An experimental evaluation; part ii, graph coloring and number partitioning. Operations Research **39**(3), 378–406 (1991)
153. Jones, D.: A taxonomy of global optimization methods based on response surfaces. Journal of Global Optimization **21**(4), 345–383 (2001)
154. Jones, T., Forrest, S.: Fitness distance correlation as a measure of problem difficulty for genetic algorithms. In: Proceedings of the Sixth International Conference on Genetic Algorithms Table of Contents, pp. 184–192 (1995)
155. Jong, K.A.D., Potter, M.A., Spears, W.M.: Using problem generators to explore the effects of epistasis. In: T. Bäck (ed.) Proceedings of the Seventh intl. conference on genetic algorithms. Morgan Kaufmann, San Francisco (2007)
156. Kauffman, S.A., Levin, S.: Towards a general theory of adaptive walks on rugged landscapes. Journal of Theoretical Biology **128**, 11–45 (1987)
157. Kautz, H., Horvitz, E., Ruan, Y., Gomes, C., Selman, B.: Dynamic restart policies. In: Eighteenth national conference on Artificial intelligence, pp. 674–681. American Association for Artificial Intelligence, Menlo Park, CA (2002)
158. Kenney, J.F., Keeping, E.S.: Mathematics of Statistics, 2nd edn. Van Nostrand, Princeton, NJ (1951)
159. Kernighan, B., Lin, S.: An efficient heuristic procedure for partitioning graphs. Bell Systems Technical Journal **49**, 291–307 (1970)
160. Kincaid, R.K., Laba1, K.E.: Reactive tabu search and sensor selection in active structural acoustic control problems. Journal of Heuristics **4**(3), 199–220 (1998)
161. Kirkpatrick, S., Jr Gelatt, C.D., Vecchi, M.P.: Optimization by simulated annealing. Science **220**, 671–680 (1983)
162. Kirkpatrick, S., Selman, B.: Critical behavior in the satisfiability of random boolean expressions. Science **264**, 1297–1301 (1994)
163. Kohavi, R., John, G.: Automatic parameter selection by minimizing estimated error. Machine Learning, pp. 304–312 (1995)
164. Krasnogor, N., Smith, J.: A tutorial for competent memetic algorithms: Model, taxonomy, and design issues. IEEE Transactions on Evolutionary Computation **9**(5), 474–488 (2005)
165. Kuhn, T.S.: The Structure of Scientific Revolutions, 2nd edn. University of Chicago Press, Chicago, IL (1970)
166. Laarhoven, P.J.M., Aarts, E.H.L. (eds.): Simulated annealing: Theory and applications. Kluwer, Norwell, MA (1987)
167. Lagoudakis, M., Littman, M.: Algorithm selection using reinforcement learning. In: Proceedings of the Seventeenth International Conference on Machine Learning, pp. 511–518 (2000)
168. Lagoudakis, M., Littman, M.: Learning to select branching rules in the DPLL procedure for satisfiability. LICS 2001 Workshop on Theory and Applications of Satisfiability Testing (SAT 2001) (2001)
169. Lagoudakis, M., Parr, R.: Least-squares policy iteration. Journal of Machine Learning Research **4**(6), 1107–1149 (2004)
170. Levy, A., Montalvo, A.: The tunneling algorithm for the global minimization of functions. SIAM Journal on Scientific and Statistical Computing **6**, 15–29 (1985)
171. Leyton-Brown, K., Nudelman, E., Shoham, Y.: Learning the empirical hardness of optimization problems: The case of combinatorial auctions. Proceedings of CP02 **384** (2002)

References 189

172. Lin, S.: Computer solutions of the travelling salesman problems. Bell System Technical Journal **44**(10), 2245–69 (1965)
173. Lourenco, H.: Job-shop scheduling: Computational study of local search and large-step optimization methods. European Journal of Operational Research **83**, 347–364 (1995)
174. Lourenco, H.R., Martin, O.C., Stutzle, T.: Iterated local search. In: F. Glover, G. Kochenberger (eds.) Handbook of Metaheuristics, pp. 321–353. Springer, Berlin (2003)
175. Luby, M., Sinclair, A., Zuckerman, D.: Optimal speedup of las vegas algorithms. Information Processing Letters **47**(4), 173,180 (1993)
176. Maier, D., Salveter, S.C.: Hysterical b-trees. Information Processing Letters **12**(4), 199–202 (1981)
177. Maron, O., Moore, A.W.: The racing algorithm: Model selection for lazy learners. Artificial Intelligence Review **11**(1–5), 193–225 (1997). URL citeseer.ist.psu.edu/maron97racing.html
178. Martin, O., Otto, S.W., Felten, E.W.: Large-step Markov chains for the traveling salesman problem. Complex Systems **5**:3, 299 (1991)
179. Martin, O., Otto, S.W., Felten, E.W.: Large-step Markov chains for the tsp incorporating local search heuristics. Operation Research Letters **11**, 219–224 (1992)
180. Martin, O.C., Otto, S.W.: Combining simulated annealing with local search heuristics. Annals of Operations Research **63**, 57–76 (1996)
181. McAllester, D., Selman, B., Kautz, H.: Evidence for invariants in local search. In: Proceedings of the national conference on Artificial Intelligence, 14, pp. 321–326. Wiley, New York, NY (1997)
182. McGeoch, C.C.: Toward an experimental method for algorithm simulation. INFORMS Journal on Computing **8**(1), 1–28 (1996)
183. Merz, P., Freisleben, B.: Fitness landscape analysis and memetic algorithms for the quadratic assignment problem. IEEE Transactions on Evolutionary Computation **4**(4), 337–352 (2000)
184. Metropolis, N., Rosenbluth, A.N., Rosenbluth, M.N., Teller, A.H.T.E.: Equation of state calculation by fast computing machines. Journal of Chemical Physics **21**(6), 10871,092 (1953)
185. Minton, S., Johnston, M., Philips, A., Laird, P.: Minimizing conflicts: A heuristic repair method for constraint satisfaction and scheduling problems. Artificial Intelligence **58**(1–3), 161–205 (1992)
186. Mitchell, D., Selman, B., Levesque, H.: Hard and easy distributions of SAT problems. In: Proceedings of the Tenth National Conference on Artificial Intelligence (AAAI-92), pp. 459–465. AAAI Press San Jose, CA (1992)
187. Mitra, D., Romeo, F., Sangiovanni-Vincentelli, A.: Convergence and finite-time behavior of simulated annealing. Advances in Applied Probability **18**(3), 747–771 (1986)
188. Moore, A.W., Schneider, J.: Memory-based stochastic optimization. In: D.S. Touretzky, M.C. Mozer, M.E. Hasselmo (eds.) Advances in Neural Information Processing Systems, vol. 8, pp. 1066–1072. The MIT Press, Cambridge, MA (1996). URL citeseer.ist.psu.edu/moore95memorybased.html
189. van Moorsel, A., Wolter, K.: Analysis and algorithms for restart (2004). URL citeseer.ist.psu.edu/vanmoorsel04analysis.html
190. Morris, P.: The breakout method for escaping from local minima. In: Proceedings of the National Conference on Artificial Intelligence, 11, p. 40. Wiley, New York, NY (1993)
191. Moscato, P.: On evolution, search, optimization, genetic algorithms and martial arts: Towards memetic algorithms. Caltech Concurrent Computation Program, C3P Report **826** (1989)
192. Mühlenbein, H., Paa, G.: From recombination of genes to the estimation of distributions i. binary parameters. In: A. Eiben, T. Bäck, M. Shoenauer, H. Schwefel (eds.) Parallel Problem Solving from Nature, PPSN IV, p. 178187 (1996)
193. Muller, S., Schraudolph, N., Koumoutsakos, P.: Step size adaptation in evolution strategies using reinforcementlearning. Proceedings of the 2002 Congress on Evolutionary Computation, 2002. CEC'02. **1**, 151–156 (2002)
194. Nahar, S., Sahni, S., Shragowitz, E.: Experiments with simulated annealing. In: DAC'85: Proceedings of the Twenty-second ACM/IEEE conference on Design automation, pp. 748–752. ACM Press, New York, NY (1985). DOI http://doi.acm.org/10.1145/317825.317977

195. Nahar, S., Sahni, S., Shragowitz, E.: Simulated annealing and combinatorial optimization. In: DAC'86: Proceedings of the Twenty-third ACM/IEEE Conference on Design Automation, pp. 293–299. IEEE Press, Piscataway, NJ (1986)
196. Nareyek, A.: Choosing search heuristics by non-stationary reinforcement learning. In: Metaheuristics: Computer Decision-Making, pp. 523–544. Kluwer, Norwell, MA (2004)
197. Nelson, W.: Applied Life Data Analysis. Wiley, New York, NY (1982)
198. Nonobe, K., Ibaraki, T.: A tabu search approach for the constraint satisfaction problem as a general problem solver. European Journal of Operational Research **106**, 599–623 (1998)
199. Nudelm, E., Leyton-Brown, K., Hoos, H., Devk, A., Shoham Y.: Understanding Random SAT: Beyond the Clauses-to-Variables Ratio. Principles and Practice of Constraint Programming–CP 2004: Tenth International Conference, CP 2004, Toronto, Canada, September 27–October 1, 2004: Proceedings (2004)
200. Oblow, E.: Pt: A stochastic tunneling algorithm for global optimization. Journal of Global Optimization **20**(2), 191–208 (2001)
201. Osman, I.H.: Metastrategy simulated annealing and tabu search algorithms for the vehicle routing problem. Annals of Operations Research **41**(1–4), 421–451 (1993)
202. Osman, I.H., Wassan, N.A.: A reactive tabu search meta-heuristic for the vehicle routing problem with back-hauls. Journal of Scheduling **5**(4), 287–305 (2002)
203. Osuna, E., Freund, R., Girosi, F.: Support vector machines: Training and applications. Tech. Rep. AIM-1602, MIT Artificial Intelligence Laboratory and Center for Biological and Computational Learning (1997)
204. Overmars, M.H.: Searching in the past ii: General transforms. Tech. Rep., Department of Computer Science, University of Utrecht, Utrecht, The Netherlands (1981)
205. Papadimitriou, C.H., Steiglitz, K.: Combinatorial Optimization, Algorithms and Complexity. Prentice-Hall, NJ (1982)
206. Parkes, A.J.: Clustering at the phase transition. In: AAAI/IAAI, pp. 340–345 (1997). URL citeseer.ist.psu.edu/parkes97clustering.html
207. Parr, R., Painter-Wakefield, C., Li, L., Littman, M.: Analyzing feature generation for value-function approximation. In: Proceedings of the Twenty-fourth International Conference on Machine Learning, pp. 737–744 (2007)
208. Patterson, D., Kautz, H.: Auto-walksat: A self-tuning implementation of walk-sat. Electronic Notes in Discrete Mathematics (ENDM) (2001)
209. Pelikan, M., Goldberg, D., Lobo, F.: A survey of optimization by building and using probabilistic models. Computational Optimization and Applications **21**(1), 5–20 (2002)
210. Pelta, D., Sancho-Royo, A., Cruz, C., Verdegay, J.: Using memory and fuzzy rules in a cooperative multi-thread strategy for optimization. Information Sciences **176**(13), 1849–1868 (2006)
211. Pincus, M.: A monte carlo method for the approximate solution of certain types of constrained optimization problems. Operations Research **18**(6), 1225–1228 (1970)
212. Prais, M., Ribeiro, C.C.: Reactive grasp: An application to a matrix decomposition problem in tdma traffic assignment. Informs Journal on Computing **12**(3), 164–176 (2000)
213. Quinlan, J.R.: Combining instance-based and model-based learning. In: Proceedings of the Tenth International Conference on Machine Learning, pp. 236–243. Morgan Kaufmann, Amherst, MA (1993). URL citeseer.ist.psu.edu/quinlan93combining.html
214. Rechenberg, I.: Evolutionsstrategie. Frommann-Holzboog (1973)
215. (reviewer), M.S.: The Wisdom of Crowds. American Journal of Physics **75**, 190 (2007)
216. Robbins, H., Monro, S.: A stochastic approximation method. The Annals of Mathematical Statistics **22**, 400–407 (1951)
217. Rochat, Y., Taillard, E.: Probabilistic diversification and intensification in local search for vehicle routing. Journal of Heuristics **1**(1), 147–167 (1995)
218. Ruan, Y., Horvitz, E., Kautz, H.: Restart policies with dependence among runs: A dynamic programming approach. CP (2002). URL citeseer.ist.psu.edu/ruan02restart.html
219. Rubinstein, R.: Optimization of computer simulation models with rare events. European Journal of Operations Research **99**, 89–112 (1997). URL citeseer.ist.psu.edu/rubinstein96optimization.html

References

220. Rubinstein, R.: The cross-entropy method for combinatorial and continuous optimization. Methodology and Computing in Applied Probability **1**(2), 127–190 (1999)
221. Ryan, J., Bailey, T., Moore, J., Carlton, W.: Reactive tabu search in unmanned aerial reconnaissance simulations. In: Proceedings of the Thirtieth Conference on Winter Simulation, pp. 873–880 (1998)
222. Sarnak, N., Tarjan, R.E.: Planar point location using persistent search trees. Communications of the ACM **29**(7), 669–679 (1986)
223. Schreiber, G.R., Martin, O.C.: Cut size statistics of graph bisection heuristics. SIAM Journal of Optimization **10**(1), 231–251 (1999)
224. Schuurmans, D., Southey, F.: Local search characteristics of incomplete sat procedures. Artificial Intelligence **132**(2), 121–150 (2001). DOI http://dx.doi.org/10.1016/S0004-3702(01)00151-5
225. Schuurmans, D., Southey, F., Holte, R.: The exponentiated subgradient algorithm for heuristic boolean programming. In: Proceedings of the International Joint Conference on Artificial Intelligence, vol. 17, pp. 334–341. Lawrence Erlbaum associates, (2001)
226. Schwefel, H.: Numerical Optimization of Computer Models. Wiley, New York, NY (1981)
227. Selman, B., Kautz, H.: Domain-independent extensions to GSAT: Solving large structured satisfiability problems. In: Proceedings of IJCAI-93, pp. 290–295 (1993)
228. Selman, B., Kautz, H.: An empirical study of greedy local search for satisfiability testing. In: Proceedings of the Eleventh National Conference on Artificial Intelligence (AAAI-93). Washington, DC (1993)
229. Selman, B., Kautz, H., Cohen, B.: Noise strategies for improving local search. In: Proceedings of the National Conference on Artificial Intelligence, vol. 12. Wiley, New York, NY (1994)
230. Selman, B., Kautz, H., Cohen, B.: Local search strategies for satisfiability testing. In: M. Trick, D.S. Johnson (eds.) Proceedings of the Second DIMACS Algorithm Implementation Challenge on Cliques, Coloring and Satisfiability, No. 26 in DIMACS Series on Discrete Mathematics and Theoretical Computer Science, pp. 521–531 (1996)
231. Selman, B., Levesque, H., Mitchell, D.: A new method for solving hard satisfiability problems. In: Proceedings of the Tenth National Conference on Artificial Intelligence (AAAI-92), pp. 440–446. AAAI Press San Jose, CA (1992)
232. Selman, B., Mitchell, D.G., Levesque, H.J.: Generating hard satisfiability problems. Artificial Intelligence **81**(1-2), 17–29 (1996). DOI http://dx.doi.org/10.1016/0004-3702(95)00045-3
233. Siarry, P., Berthiau, G., Durdin, F., Haussy, J.: Enhanced simulated annealing for globally minimizing functions of many-continuous variables. ACM Transactions on Mathematical Software **23**(2), 209–228 (1997). DOI http://doi.acm.org/10.1145/264029.264043
234. Singer, J., Gent, I., Smaill, A.: Backbone Fragility and the Local Search Cost Peak. Journal of Artificial Intelligence Research **12**, 235–270 (2000). URL citeseer.ist.psu.edu/singer00backbone.html
235. Khanna, S., Motwani, R., Sudan, M., Vazirani, U.: On syntactic versus computational views of approximability. In: Proceedings of the Thirty-fifth Annual IEEE Symposium on Foundations of Computer Science, pp. 819–836 (1994)
236. Smith, B.: Phase transition and the mushy region in constraint satisfaction problems. In: Proceedings of the Eleventh European Conference on Artificial Intelligence, pp. 100–104 (1994)
237. Smith, R.E., Smith, J.E.: New methods for tunable random landscapes. In: W.N. Martin, W.M. Spears (eds.) Foundations of Genetic Algorithms 6. Morgan Kaufmann, Los Altos, CA (2001)
238. Smola, A.J., Schölkopf, B.: A tutorial on support vector regression. Tech. Rep. NeuroCOLT NC-TR-98-030, Royal Holloway College, University of London, UK (1998)
239. Southey, F.: Theory and Applications of Satisfiability Testing, chap. Constraint Metrics for Local Search, pp. 269–281. Springer, Berlin (2005)
240. de Souza, P.S., Talukdar, S.N.: Asynchronous organizations for multi-algorithm problems. In: SAC'93: Proceedings of the 1993 ACM/SIGAPP Symposium on Applied Computing, pp. 286–293. ACM Press, New York, NY (1993). DOI http://doi.acm.org/10.1145/162754.162902

241. Steiglitz, K., Weiner, P.: Algorithms for computer solution of the traveling salesman problem. In: Proceedings of the Sixth Allerton Conference on Circuit and System Theory, Urbana, IL, pp. 814–821. IEEE, New York, NY (1968)
242. Streeter, M.J., Smith, S.: A simple distribution-free approach to the max k-armed bandit problem. In: Proceedings of the Twelfth International Conference on Principles and Practice of Constraint Programming (CP 2006) (2006)
243. Streeter, M.J., Smith, S.F.: An asymptotically optimal algorithm for the max k-armed bandit problem. In: AAAI (2006)
244. Strenski, P.N., Kirkpatrick, S.: Analysis of finite length annealing schedules. Algorithmica **6**, 346–366 (1991)
245. Stutzle, T., Hoos, H., et al.: MAX-MIN ant system. Future Generation Computer Systems **16**(8), 889–914 (2000)
246. Surowiecki, J.: The wisdom of crowds: Why the many are smarter than the few and how collective wisdom shapes business, economies, societies, and nations. Doubleday (2004)
247. Sutton, R., Barto, A.: Introduction to Reinforcement Learning. MIT Press Cambridge, MA (1998)
248. Taillard, E.: Robust taboo search for the quadratic assignment problem. Parallel Computing **17**, 443–455 (1991)
249. Talbi, E.G.: Parallel Combinatorial Optimization. Wiley, New York, NY (2006)
250. Tarjan, R.E.: Updating a balanced search tree in o(1) rotations. Information Processing Letters **16**, 253–257 (1983)
251. Tompkins, D., Hoos, H.: Warped landscapes and random acts of SAT solving. In: Proceedings of the Eighth International Symposium on Artificial Intelligence and Mathematics (ISAIM-04) (2004)
252. Tompkins, F.H.D., Hoos, H.: Scaling and probabilistic smoothing: Efficient dynamic local search for sat. In: Proceedings of the Principles and Practice of Constraint Programming – CP 2002: Eighth International Conference, CP 2002, Ithaca, NY, September 9–13, LNCS, vol. 2470, pp. 233–248. Springer, Berlin (2002)
253. Toune, S., Fudo, H., Genji, T., Fukuyama, Y., Nakanishi, Y.: Comparative study of modern heuristic algorithms to service restoration in distribution systems. IEEE Transactions on Power Delivery **17**(1), 173–181 (2002)
254. Valentini, G., Masulli, F.: Ensembles of learning machines. In: M. Marinaro, R. Tagliaferri (eds.) Neural Nets WIRN Vietri-02, LNCS. Springer, Heidelberg, Germany (2002). URL citeseer.ist.psu.edu/valentini02ensembles.html
255. Vapnik, V., Chervonenkis, A.J.: On the uniform convergence of relative frequencies of events to their probabilities. Theory of Probability and Its Applications **16**, 264–280 (1971)
256. Vapnik, V.N.: The Nature of Statistical Learning Theory. Springer, Berlin (1995)
257. Verhoeven, M., Aarts, E.: Parallel local search. Journal of Heuristics **1**(1), 43–65 (1995)
258. Vossen, T., Verhoeven, M., ten Eikelder, H., Aarts, E.: A quantitative analysis of iterated local search. Computing Science Reports 95/06, Department of Computing Science, Eindhoven University of Technology, Eindhoven, The Netherlands (1995)
259. Voudouris, C., Tsang, E.: The tunneling algorithm for partial CSPs and combinatorial optimization problems. Tech. Rep. CSM-213 (1994). URL citeseer.ist.psu.edu/voudouris94tunneling.html
260. Voudouris, C., Tsang, E.: Partial constraint satisfaction problems and guided local search. In: Proceedings of Second International Conference on Practical Application of Constraint Technology (PACT 96), London, pp. 337–356 (1996)
261. Voudouris, C., Tsang, E.: Guided local search and its application to the traveling salesman problem. European Journal of Operational Research **113**, 469–499 (1999)
262. Wah, B., Wu, Z.: Penalty formulations and trap-avoidance strategies for solving hard satisfiability problems. Journal of Computer Science and Technology **20**(1), 3–17 (2005)
263. Wang, C., Tsang, E.: Solving constraint satisfaction problems using neural networks. In: Proceedings of the Second International Conference on Artificial Neural Networks, pp. 295–299 (1991)

264. Watson, J.P., Beck, J.C., Howe, A.E., Whitley, L.D.: Problem difficulty for tabu search in job-shop scheduling. Artificial Intelligence **143**(2), 189–217 (2003). DOI http://dx.doi.org/10.1016/S0004-3702(02)00363-6
265. Weinberger, E.: Correlated and uncorrelated fitness landscapes and how to tell the difference. Biologial Cybernetics **63**, 325–336 (1990)
266. White, S.: Concepts of scale in simulated annealing. In: AIP Conference Proceedings, vol. 122, pp. 261–270 (1984)
267. Whitley, D., Gordon, V., Mathias, K.: Lamarckian Evolution, The Baldwin Effect and Function Optimization. In: Parallel Problem Solving from Nature–PPSN III: International Conference on Evolutionary Computation, Jerusalem, Israel. Springer, Berlin (1994)
268. Winter, T., Zimmermann, U.: Real-time dispatch of trams in storage yards. Annals of Operations Research **96**, 287–315 (2000). URL citeseer.ist.psu.edu/winter00realtime.html
269. Wolpert, D.H., Macready, W.G.: No free lunch theorems for optimization. IEEE Transactions on Evolutionary Computation **1**(1), 67–82 (1997). URL citeseer.ist.psu.edu/wolpert96no.html
270. Woodruff, D.L., Zemel, E.: Hashing vectors for tabu search. Annals of Operations Research **41**, 123–138 (1993)
271. Youssef, S., Elliman, D.: Reactive Prohibition-Based Ant Colony Optimization (RPACO): A New Parallel Architecture for Constrained Clique Sub-Graphs. In Proceedings of the Sixteenth IEEE International Conference on Tools with Artificial Intelligence (ICTAI'04)0 pp. 63–71 (2004)
272. Zhang, Q.: On stability of fixed points of limit models of univariate marginal distribution algorithm and factorized distribution algorithm. IEEE Transactions on Evolutionary Computation **8**(1), 80–93 (2004)
273. Zhang, Q., Muhlenbein, H.: On the convergence of a class of estimation of distribution algorithms. IEEE Transactions on Evolutionary Computation **8**(2), 127–136 (2004)
274. Zhang, W., Dietterich, T.: A reinforcement learning approach to job-shop scheduling. In: Proceedings of the Fourteenth International Joint Conference on Artificial Intelligence **1114** (1995)
275. Zhang, W., Dietterich, T.: High-performance job-shop scheduling with a time-delay TD (λ) network. Advances in Neural Information Processing Systems **8**, 1024–1030 (1996)
276. Zlochin, M., Birattari, M., Meuleau, N., Dorigo, M.: Model-based search for combinatorial optimization. Annals of Operations Research **131**, 373–395 (2004)

Index

k-armed bandit problem, 141

attraction basins and escape, 41
autocorrelation, 166

Baldwin effect, 156
breakout technique, 60

Chernoff inequality, 142
churn, 161
clauseweighting, 60
correlation coefficient, 104
correlation length, 167
correlation ratio, 104
cross-entropy, 75

design of experiments, 79
DLS, 62
dynamic programming
    Bellman equation, 120
dynamical systems, 37

entropy, 103, 105
    conditional, 106
epistasis, 168
ESG, 62
evolution strategies, 152

feature selection, 102
fixed tabu search, 37

genetic algorithms, 152
GLS, 63
gradient descent, 11
GRASP, 76
    Reactive —, 77
GSAT, 59

hashing and fingerprinting, 54

iterated local search, 18

kriging, 79

Lamarckian evolution, 154, 156
landscape correlation function, 166
local search
    attraction basin, 11
    big valley, 12
    local minimum, 10
    local optimum, 10
    perturbation, 9
    search trajectory, 10
Locally Weighted Regression
    Bayesian, 89

Markov decision process, 118
Markov processes, 26
memetic algorithms, 156
MIMIC, 73
model, 69
model-based optimization, 69
mutual information, 103, 106

neuro-dynamic programming, 118
NK landscape model, 168
non–oblivious local search, 66

off-line configuration, 145

P2P, 160
peer-to-peer, 160
penalty-based search, 59
perceptron, 93
persistent dynamic sets, 54

portfolio, 129
probabilistic tabu search, 38
prohibition and diversification, 40

racing, 141
reactive tabu search, 38, 49
regression
 ridge —, 87
reinforcement learning, 14, 117
response surfaces, 79
restart, 135
robust tabu search, 38
RSAPS, 62

scatter search and path relinking, 159

search history storage, 52
simulated annealing, 26
 adaptive, 33
 nonmonotonic cooling schedules, 31
 phase transitions, 30
strict tabu search, 37

tabu search, 35
Teams of solvers, 151
Tichonov regularization, 87

variable neighborhood search, 14

WGSAT, 61

Early Titles in
OPERATIONS RESEARCH/COMPUTER SCIENCE INTERFACES

Greenberg / *A Computer-Assisted Analysis System for Mathematical Programming Models and Solutions: A User's Guide for ANALYZE*
Greenberg / *Modeling by Object-Driven Linear Elemental Relations: A Users Guide for MODLER*
Brown & Scherer / *Intelligent Scheduling Systems*
Nash & Sofer / *The Impact of Emerging Technologies on Computer Science & Operations Research*
Barth / *Logic-Based 0-1 Constraint Programming*
Jones / *Visualization and Optimization*
Barr, Helgason & Kennington / *Interfaces in Computer Science & Operations Research: Advances in Metaheuristics, Optimization, & Stochastic Modeling Technologies*
Ellacott, Mason & Anderson / *Mathematics of Neural Networks: Models, Algorithms & Applications*
Woodruff / *Advances in Computational & Stochastic Optimization, Logic Programming, and Heuristic Search*
Klein / *Scheduling of Resource-Constrained Projects*